AF552716

Impact of Global Climate Change on Earth Ecosystems

Impact of Global Climate Change on Earth Ecosystems

Editors
Prof. D.R. Khanna
Prof. A.K. Chopra
Dr. Gagan Matta
Dr. Vikas Singh
Dr. Rakesh Bhutiani

Associate Editor
Dr. Chakresh Pathak

2013
BIOTECH BOOKS®

ISBN 978-81-7622-273-0

Published by: **BIOTECH BOOKS®**
4762-63/23, Ansari Road, Darya Ganj,
New Delhi - 110 002
Phone: +91-011-23262132
E-mail: biotechbooks@yahoo.co.in

Printed at:**Chawla Offset Printers**
Delhi - 110 052

PRINTED IN INDIA

Preface

There is growing consensus among the scientific community that global warming caused by increased concentration of greenhouse gases in the atmosphere is one of the most serious environmental problems facing the world today. It is estimated that global mean surface temperature could rise over 6 °C by 2100. In addition to global warming, increased greenhouse gas concentrations may increase the occurrence of precipitation extremes: greater precipitation is expected in already-wet areas and increased drought in already-dry areas. Furthermore, widespread expansion of industry and agricultural activities may increase atmospheric nitrogen deposition to unprecedented levels, which will modify climate change impacts. Climate change is also expected to increase the severity and frequency of wildfire, floods, and pest and pathogen attacks. These global environmental changes will pose serious consequences for the overall functioning of terrestrial ecosystems, particularly for agriculture and forestry.

Various ecosystems togetherly form the earth supporting life on the earth; it serves as a platform for perturbations to biogeochemical processes of importance to global climate change. In view of changing climate, it is vital that ecosystems health is maintained because a healthy ecosystem is able to sustain physical, chemical and biological functions, and recover following perturbations, due to inherent resilience. A healthy ecosystem enhances plant productivity, promotes plant, animal and human health, maintains water and air quality, supports a diverse community of organisms and resists stresses of human impact and climatic perturbations, so resists environmental degradation. A healthy ecosystem is a complex dynamic living resource that is resilient as a result of its capacity for self-organisation.

With contributions from internationally renowned experts, this book will be a great knowledge resource on the impacts of global climate change and earth ecosystems. We believe that book will help students and researchers in climate change, ecosystems and environmental sciences, as well as policy makers and industry stakeholders involved in natural resource management, agricultural development and climate change mitigation. The compiled information is expected to generate stimulating discussions among scientists and will assist in formulating research aiming to tackle knowledge gaps identified by the contributors. We thank the authors for their contribution to this volume and appreciate their diligence in responding to reviewers' comments, thereby ensuring high standards. Book is the part of World Congress of Man and Nature (WCMANU) "Global Climate Change and Biodiversity Conservation" held at Department of Zoology and Environmental Science, Gurukula Kangri University, Haridwar, India from 11th to 13th November 2011. A substantial proportion of the material is original, and has been prepared by authors specifically for this book as a part of WCMANU – 2011.

Hopefully research work published in this book will serve as a supplement in the field of global climate change. We thank all the contributors who has devoted their precious time in contributing their papers for this volume. Lastly but not least, the thanks are also due to Biotech Publishers, New Delhi in bringing out this volume in time with nice presentation.

Editors

Contents

2013, Impact of Global Climate Change on Earth Ecosystems *Pages* **1–12**
Editors: **D.R. Khanna, A.K. Chopra, Gagan Matta, Vikas Singh & Rakesh Bhutiani**
Published by: **BIOTECH BOOKS, NEW DELHI**

Chapter 1
Stakeholders' Network Governance and Regional Sustainable Development

Gislaine Garbelini and Patrícia Almeida Ashley
LATEC – Laboratory Technology, Business Management and Environment
Universidade Federal Fluminense - Niterói, Rio de Janeiro/Brasil

This article is based on a Master research that analyzed stakeholders' network governance policies and practices, designed to promote regional sustainable development. Given the necessary involvement of different stakeholders in economic, environmental and social fields in order to obtain a more balanced and fairer society, a lack of stakeholder engagement in the organizational sphere may hinder the implementation of a stakeholder governance model that promotes sustainable development.

In order to investigate this assumption, Petrobras' Environmental Center of Excellence (Ceap Amazônia) was used as a case study. This is a qualitative and exploratory study, involving bibliographic, documental and field research.

The study concludes that traditional methods of corporate governance have not been effective in achieving complex goals such as regional sustainable development. Hence the need to establish new models of governance that involve stakeholders engagement aligned with sustainability values. The theoretical framework is based on the following themes: stakeholder theory, networks, corporate governance and sustainability indicators. These themes support the conceptual analytical model known as the GovRedeS model used in the case study. Its application showed new aspects that have been incorporated into a revised version of the GovRedeS.

The model was shown to be appropriate and consistent in investigating the Ceap Amazônia network, and may be used for other investigations, considering that necessary adjustments should be made to the organizational structures and cultures of each network and their political, cultural, environmental and social environment.

Keywords: *Governance, Networks, Stakeholders, Sustainable development.*

Introduction

Organisations continuously need internal adaptations, such as management changes, restructuring, new technologies of production and logistics, among others. However, changes in external scenario compromise even more the performance and the existence itself of every type of organisation. With all technological and social changes, it is obvious the companies need to adapt a new way of managing, producing, consuming and relating to each other. Those themes have been largely discussed by specialists from varied areas of knowledge and require organisations to expand their relationships beyond investors and employees, evolving the inner management to a management which understands the importance of stakeholders in all dimensions of business.

The idea of what is considered responsibility of the company has also been changing at an international level, as stressed by Ashley (2006) when affirming the classic shareholder view represents a company focused only in answer the interest of its shareholders. On the other hand, the stakeholders view considers there are many interested parties related to organisational existence.

In the context of changes in the way business is conducted, the network theme is also addressed in many organisational studies, mainly those focused on competitive advantages and on new production relationships which may be generated by networks. These change gains make companies improve their performance, reduce costs and increase profits. Thus, most recent researches on stakeholders' network are focused on investigating transactional relationships they can produce, and not on transformational results that can generate values at an organisational and territorial level. So, this article, based on a Master research performed at Universidade Federal Fluminense, has the purpose of presenting a diagnosis of stakeholder´s network governance policies and practices for sustainable development.

Analysis of the change aspects in corporate governance is a way to detect possible difficulties and opportunities for implementing enterprise policies aligned with sustainability. In this regard, organisations must be understood in a wider way, as proposed by Ashley (2006, p. 110) when affirming that the company must be seen "as a network of relationships between stakeholders associated to the business contextualised in time and space, and before ethical challenges and in an effort to bring enterprise speech and practice together". This way, organisations are inserted in network of relationships relevant to development or their business, whether in an economic, social or environmental dimension.

If the lack of stakeholders engagement in organisational sphere may hinder the implementation of sustainable development governance model, the question is: What are the shareholder´s network governance practices and policies which contribute to promoting a regional sustainable development?

The first supposition is the performance of organisational networks for common goal of regional sustainable development requires good governance practices and policies from stakeholders' network.

Theoretical Basis

The theoretical background guiding the proposition of GovRedeS Model developed in this research was based on the following themes: theories of stakeholders, networks, corporate governance and sustainability indicators. For the purposes of this article, the main points used in theories of stakeholders' network and corporative governance will be presented very concisely.

The new scenarios, on which organisations operate, cause deep changes in the way companies are managed and they deal with its stakeholders. However, as stated by Alves (2001), its purpose is still producing goods or provide services in the most efficient and economical way. In this regard, in spite of still looking for profit for their stakeholders, companies started to manage wealth in a wider sense, such as, for instance, producing knowledge and technology, creating jobs, infrastructure, and also creating a wealth of intangible values, such as dignity at work, promotion of gender equality, sustainable development, among others. The company becomes one of the protagonists of the development of a region or country, with direct political and economic implications in society.

So, the company's role as wealth creator evokes this question: is it feasible to define a social function to companies and, if so, is it possible or desirable to establish certain obligations with society? Regarding this, Moore (1999) proposes the discussion about stakeholders takes in consideration that the companies have, as a goal, not only create wealth, but also distribute it fairly, considering their personal and interorganisational relationships with environment and society.

This comprehensiveness in concept of wealth generation involves a new producing way with values which overcome the search for profit maximisation. According to Alves (2001), a corporate change in attitude now that companies have been careful of values makes them more similar to citizen companies and it is necessary to analyse them from the viewpoint of corporate governance (CG) phenomenon. In his opinion, if corporate citizenship and corporate governance evolve in business environment, companies may perform their role widely, in other words, "generate wealth and value to society in a sustainable basis". (Alves, 2001, p. 86).

A Brief Thought About Network Theory Applied to Organisations: Networks of Stakeholders

Before defining the delimitation of networks of stakeholders for this article, it is important to analyse some network theory. For Castells (2002), it can be defined as a set of interconnected nodes in which relationships result in disruptions of boundaries, relaxation of hierarchical structures, and integration of activities. Each network node is described by Stewart (1998) as the agents who are part of it, being individually allocated, albeit interdependently. Also in accordance with Castells (2002), the formation and performance of a network depend on two aspects: connectivity – ability to communicate without noise among their components, and coherency – reach of a common planning sharing the goals of the network and of its actors. Marteleto (2001) highlights the power relationships resulting from this new structure of non-hierarchical and spontaneous organisation must be considered, aiming to understand to the extent the dynamics of knowledge and information interferes in this process.

Marchi *et al.* (2007) quote a research performed in corporate networks from the United States in which Human and Provan (1997) classified into two groups the results obtained from companies acting in networks: transactional (promoting performance gains and resource acquisitions, reducing costs), and transformational results (promoting changes in the way of conducting, resulting in value creation). As highlighted by Marchi *et al.* (2007), although the companies mainly look for gains with cost reductions, the transformational results are those that promote innovations and make business become sustainable in the long-term. So, it is possible to imply these results must be sought by companies concerned about corporate social responsibility and about sustainable development of their business and of society.

Migliori (2009) makes a relevant distinction of networks by requesting that everyone has access to the same knowledge, it is not a network, it is an information flow. The network is formed in the movement of interrelationships, in which is possible to access the knowledge of everyone involved.

However, what would be the reasons to make organisations leave a comfort position, working individually, to operate in network with other organisations? In the process of formation of interorganisational networks, Oliver (1990 apud PERIM; ZAQUETTO FILHO, 2007) highlights six contingency factors which explain why these organisations are interested in this type of relationship: a) need: which may be legal or regulatory; b) asymmetry: potential of exerting power and control over other organisations or their resources; c) reciprocity: relationships based on cooperation, collaboration and coordination among organisations start from common goals and interests; d) efficiency: guidance focused on internal aspects; e) stability: uncertainties take organisations to establish relationships with others looking for stability, predictability, and an environment of trust; f) authenticity: increase of authenticity from the relationship with other reputable organisations in the market.

Among the advantages of organisational network structure, Wegner, Wittmann, and Dotto (2006) summarise the main benefits of networks in: learning, information and knowledge sharing, authenticity gains, status, and improvement of financial and management performance. Though there are many advantages of acting in networks, this structure faces many managing and governance problems. Cruz, Mugnaini, and Martins (2007, p. 4) highlight "the balance between little management and productivity versus much management and destruction of informal nature, which allows the interorganisational operation in a collaborative way, is the main challenge of network management". In their opinion, overcoming this conflict demands an intermediate management system conciliating the formality required for business with the flexibility and adaptability required for networks.

Volpon and Macedo Soares (2007, p. 409) highlight the trouble of network management diagnosing that the main weaknesses of the organisation they studied "were concentrated in the dimension of network management, related to the process of change management, and its little experience with social alliances, classified as real and potential weakness, respectively".

Despite the countless advantages, the problems in management reinforce the importance of the network governance theme and the need for further studies, mainly those related to transformational networks focused on complex subjects, such as environment and sustainable development.

Governance Aimed at Networks of Stakeholders: Instrument for Promotion of Sustainable Development Model

The direct relationship between good governance and sustainable development appears on ISO 26.000 document (INTERNATIONAL, 2008) stating that the performance of an organisation regarding society and environment became a critical part in measuring its performance, as a result of the recognition of the need for ensuring healthy ecosystems, social equality, and good organisational governance. (INTERNATIONAL, 2008).

Ashley and Macedo-Soares (2007, p. 7) defend that it is necessary to incorporate the "philosophy from social responsibility to governance of business-society transactions, going through all the network of stakeholders, so that the concept itself of social responsibility in business is sustainable (.) in the context of market economy".

The engagement of stakeholders is defined as the "activity performed to create opportunities of conversation between the organisation and one or more of its stakeholders, aiming to provide an information base for the organisation's decisions" (INTERNATIONAL, 2008, p. 11). In order to achieve this, it is necessary to identify the stakeholders, respect the authenticity of their interests, recognize their capabilities of influencing the organisation's destiny, and considering the stakeholders' opinions which may affect or be affected by the organisation's decisions.

Both in engagement of stakeholders and in promoting transparency, communication is a key instrument for enabling governance structures concerned with ethical and common good issues. Communication builds credibility by establishing a conversation with stakeholders and increasing the trust that their interests are considered.

So, it is implied that the relationships in network with stakeholders – involving aspects such as cooperation, communication, equality, engagement, and transparency – are fundamental to establish a proper structure for governance and corporate citizenship focused on promoting sustainable development, whether its dimension is local, regional, or national.

Method and Proposition of GovRedeS Model

From the identified correlations between theories of networks of stakeholders and corporate governance, an Analytical Conceptual Model of Governance of Networks of Stakeholders for Sustainable Development (Modelo Conceitual Analítico de Governança de Redes de Stakeholders para o Desenvolvimento Sustentável) – GovRedeS Model is proposed.

Similarly to the concept adopted by Sustainable Development Indicators (Indicadores de Desenvolvimento Sustentável) from the Brazilian Institute of Geography and Statistics (IBGE), the proposition is developed from the four dimensions of sustainable development: economic, social, environmental, and institutional.

The model is also conceptually based on the definition of transformational results proposed by Human and Provan (1997 and Marchi *et al.*, 2007). According to this definition, more than transactional results (such as performance gains, for instance), companies acting in network may obtain transformational results which promote changes in the way business are conducted, resulting in value creation.

To test the Model, an exploratory and qualitative research was performed, investigating the relationships between governance practices and policies of governance of networks of stakeholders and regional sustainable development from the variables of the Analytical Conceptual Model GovRedeS. The case study was complemented by three other investigation means: document, bibliographical and field research.

The universe of the research represents the total of actors composing the network of Petrobras' Environmental Excellence Centre in Amazon – (Centro de Exelência Ambiental Petrobras - Ceap Amazônia), namely: Petrobras (energy company, supporter of the network), universities, research centres, governments, communities, economic agents, civil society entities, investors, and other groups interested in oil activities in the region.

Since the focus of the paper is network governance, the sample selection was made by typical features among the subjects directly involved in its planning and management, in other words, individuals from specific areas in Petrobras who are, ultimately, responsible for network governance were chosen.

The choice of Ceap Amazônia was due to the fact it is a network aiming the excellence in socio-environmental sustainability of the activities of oil and gas production chain in Amazon. This way, as a goal of this research, the Centre presented thematic and structural characteristics suitable for the research goals, especially for application of GovRedeS Model.

The field research was performed through interviews with representatives from areas responsible for creation and management of Ceap Amazônia: Organisation, Management and Governance (Organização, Gestão e Governança - OGG); Safety, Environment, Energy Efficency and Health (SMES); Centro de Pesquisa e Desenvolvimento Leopoldo Miguez (Cenpes), and Institutional Communication (CI).

After the data gathering, the content analysis was performed in three steps. In the first step – pre-analysis – recordings of the interviews were listened to and transcribed, according to relevance of information to the goals of the research. The data organisation followed the order of the questions in the form, grouped in the following blocks:

1. specific question for each interviewed area, in which was possible to recover aspects of network governance in the way of the interviewee's area of activity.

After that, there were general questions aimed to answer the variables which compose, until that time, the GovRedeS Model:

1. Engagement of stakeholders;
2. Values for sustainable development;
3. Equality in distribution of roles and power;
4. Cooperation;
5. Responsibility for products and process;
6. Transparency;
7. Open question about aspects the interviewee considered relevant to network governance and which have not been addressed in the previous questions in the form.

In this step, the main problems and advantages for activity in network in Ceap Amazônia were identified, as follows:

- ☆ Problems for activity in network in Ceap Amazônia: organisational culture, mistrust about security of information, managing fragmentation, structural differences among the links.
- ☆ Advantages for activity in network in Ceap Amazônia: strengthen geopolitical position of Brazil, knowledge management, potentialise resources, maintain the talented in the region, improve reputation of the links.

From the systemisation of contents, the second step begun, in which an analytical description was performed through an in-depth reading of the material, considering the relationships among the themes addressed in the form and their respective variables. In this step, it was possible to identify unpredicted aspects in GovRedeS model which pointed to a necessity of review of the initially proposed Model.

The third step of the content analysis was the background interpretation which identified connections between the theoretical basis of this research, the GovRedeS Model, and the case study of Ceap Amazônia.

Discussion

The variables for a good governance of networks of stakeholders in the GovRedeS model were initially defined from a theoretical background. With its application in the case study of Ceap Amazônia, it was checked that it was possible to improve it adding new aspects and variables.

As highlighted by Sachs (2002), the large configuration of socio-economic and cultural diversities excludes the possibility of a general application of development strategies. In this regard, it is worth noting that other variables may be considered, according to the structures and organisational cultures from each network and its links, in addition to the contexts they act. Thus, the GovRedeS Model started being composed by eight variables:

Table 1.1

Variables	*Description*
Engagement of stakeholders	Systematised process of engagement and dialogue with interested parties regarding structure and organisational process for continuous improvement of governance of network of stakeholders, including: - involvement of interested parties in management; - respect to diversity (cultural, economic, social, environmental and gender); - formalisation of managing committees/boards; - transparent ways to manage conflicts of interest; - transparency in the system of payment for managing committees/boards; - "hearing" mechanisms of interested parties; - forums to promote permanent dialogue.
Values for sustainable development	Formalisation of values and ethic obligations aiming sustainable development and ways to disseminate it among stakeholders, such as code of ethics and communication channels.
Equality	Balance in distribution of roles among network stakeholders; level of power distribution; opportunity to participate and express in a democratic and egalitarian way.
Cooperation	Relationships based on cooperation and collaboration in search of common goals and interests, including: coherency to reach a common planning which shares the goals of the network and of its links; organicity and cohesion (unity) among the links of network, and among these and the society; connectivity (capacity to communicate without noise among the components).
Responsibility for products and process	Valuation and assessment of economic, social, and environmental impacts from process and products generated by the agents of the network.
Transparency	Strategies, channels, and frequency of communications with stakeholders. Transparency in communication processes and relationships with stakeholders. Transparency in accountability and information. Provide, in a clear and straightforward way, relevant information for interested parties, such as social balance sheets, for instance. Democratic environment to look for collective solutions for problems.
Negotiation and Management of conflicts	Relationships able to generate good results for every link of the network (win-win relationships), including: capacity to negotiate fairly for all parties; protection and strengthening of weaker links; formalisation of discussion forums with participation of stakeholders' representatives; formalised instances to manage conflicts.
Knowledge and social capital management	Production and sharing of knowledge and social capitals, including: recording, systematisation and dissemination of knowledge and experiences; considering of potentialities resulting from social capital of network links, respect to traditional knowledge of communities.

With insertion of two new variables ("Negotiation and Management of conflicts" and "Knowledge and social capital management") the Model has the following configuration:

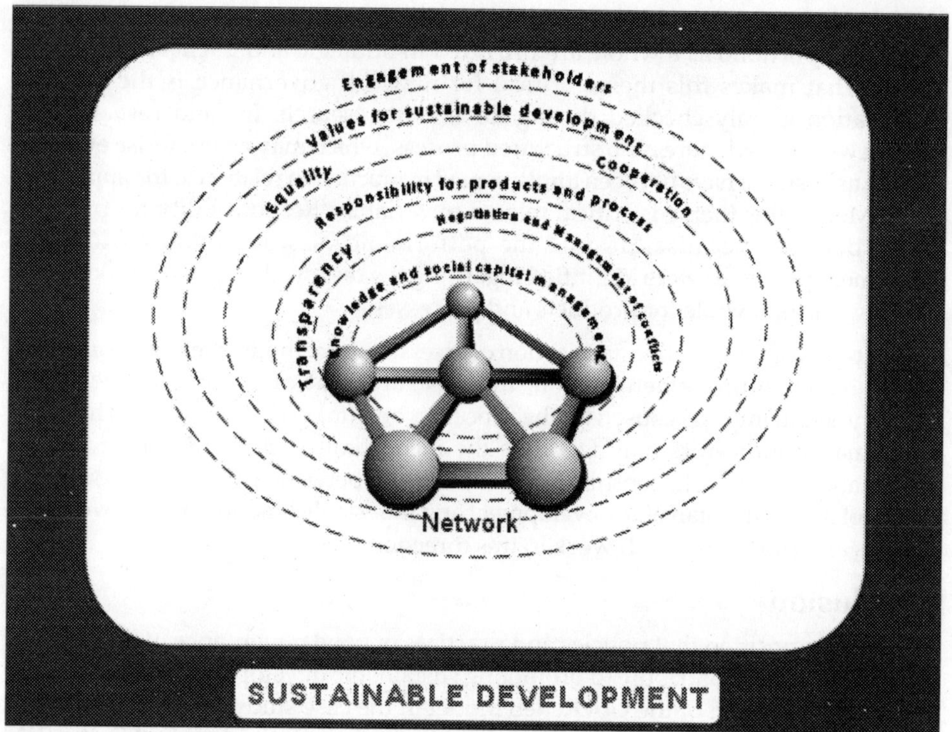

Figure 1.1

The content analysis pointed three critical themes to governance of network of stakeholders of Ceap Amazônia: stakeholders, organisational cultures, and responsibilities for products and processes.

Regarding stakeholders, it was possible to check the recognition of importance of transparency and dialogue for the collective building of solutions. However, that recognition is focused on some specific stakeholders. Thus, it is necessary to undertake in network governance ways of engagement and relationship with all stakeholders – according to the role each one has in network operation. A plan with specifies channels and frequency of communication focused on many stakeholders and the accountability to society will contribute to the promotion of transparency in network.

That way, through theoretical basis and the data collected in field research, it can be concluded that the work of identification and engagement of stakeholders shall be part of the improvement process of governance in the Centre, being able to leverage the performance of other variables in GovRedeS Model, such as "Values for Sustainable Development", "Equality", and "Transparency".

The organisational cultures theme was pointed out in the interviews, recurrently, was an internal obstacle to network governance of Ceap Amazônia, both by the

Petrobras side and some universities and research institutions. The expression is used in plural, as it refers to many cultures of the links which, together, constitute the organisational culture of the network.

In regard to impact assessment theme, assessments of performance, results and the project portfolio as a whole are provided in Strategic and Tactical Planning of Ceap. What makes this theme critical for network governance is the fact that information is only checked during document research. In field research, the interviewees stated there is no structured process, which may characterise evidence that the assessment have not been implemented in practice. In relation to the application of the Model, this theme is critical related to variable "Responsibility for products and processes" because, despite the performance assessments provided, no assessment processes were identified regarding the responsibility of the links and of the network as a whole for products and processes.

In terms of the effective application of assessment methods, the Centre may also adopt, as an improvement action, the measurement of economic, social and environmental impacts caused by the processes and links of the network. The use of performance indicators may facilitate the incorporation of results in process of governance of network, contributing to coherency between the goal of working in favour of regional sustainable development and the establishment of good governance policies and practices which work in this direction.

Conclusion

The supposition that policies and practices of good governance of networks of stakeholders may contribute to promoting sustainable development was confirmed with the application of the GovRedeS Model in the case study of Ceap Amazônia, which is proved in the critical points of governance of Ceap identified in the field research: 1. lack of stakeholder engagement; 2. aspects of organisational cultures of the supporter and its links; 3. lack of assessment processes regarding responsibility of the links and of the network as a whole for processes and products. The insertion and adaptation of variables proves the Model has necessary flexibility to be applied in other realities, since the quirks of each network and region are considered.

In relation to Ceap Amazônia, it is worth noting that the relevant aspects for a good governance of network are addressed in its Strategic and Tactical Planning, but field research identified that some of them – such as engagement of stakeholders, for instance – still need to be implemented in the practice of management of the Centre.

Among the expected benefits of the network, there is the strengthening of institutional relationship with an audience considered strategic by the supporting company, such as universities, public power and civil society organisations. However, in practice, only universities and research institutions are already counting on effective actions implemented to their engagement.

Another expected benefit is the integration with the Amazonians and their regional entities, but, according to the result of the research, civil society communities and organisations are still seen unilaterally, without direct participation in its governance. Such engagement is essential to an agreed participation able to establishing win-win relationships for the communities (which can use the results of network actions to improve the quality of life) and for the network itself (which can take advantage from social capital and traditional knowledge of the communities to reach its goals).

In this regard, the structured work of identification and engagement of stakeholders shall be part of the improvement process of governance of the Centre, being capable of leveraging other aspects of good governance, such as equality and transparency. The categorization of the press and the production chain of oil and gas in Amazon may also contribute to dissemination of important values to promote regional sustainable development.

Ceap's Planning also highlights, among the managing goals, the implementation of a simple and agile model, focused on results. The achievement of that goal has been compromised by issues associated with a historically centralising organisational culture. One of the strategies referred to in the Plan is the adoption of advanced institutional arrangements to make partnerships possible and which are adequate to the features of a specific action and to the partner institution, which shows a predisposition of Ceap to advance to new governance structures and to adapt to different realities.

This reinforces that the network is already well based, in relation to planning. In practice, it has been facing obstacles deeply rooted in organisational culture which can be overcome with the maturation of actions (the Centre has been working for a few years), and improvement of network governance and in the way its processes are assessed.

Regarding the Model, the field research proved the pertinence and relevance of its variables for the diagnosis of policies and practices of governance of network of stakeholders focused on regional sustainable development. Therefore, its use is recommended for other investigations, since the necessary adaptations to different structures and organisational cultures of each network, and the political, cultural, environmental and social contexts they act are taken in consideration.

References

Alves, Lauro Eduardo Soutello (2001). *Governança e cidadania empresarial.* {RAE, São Paulo, Vol. 41, no. 4, pp. 78-86, Oct./Dec.

Ashley, Patrícia A.; and Macedo-Soares, T. D. L. V. A. (Org.) (2006). *Ética e responsabilidade social nos negócios.* 2nd. ed. São Paulo: Saraiva.

______ (2002). *Um modelo conceitual para a incorporação da responsabilidade social à governança das relações negócio-sociedade.* Rio de Janeiro: PUC, 2002. Available at: <http://www.iag.puc-rio.br/sobre/tds/TD06_UM per cent 20MODELO per cent 20CONCEITUAL.pdf>. (Accessed: 9 Mar. 2009).

Castells, M. (2002). *A sociedade em rede.* 6th. ed. São Paulo: Paz e Terra.

Cruz, June Alisson Westarb; Mugnaini, Alexandre; and Martins, Tomas Sparano (2007). *Governança em redes como forma de promoção de qualidade de vida.* In: ENANPAD, 31., Rio de Janeiro. *Resumo dos trabalhos.* Rio de Janeiro.

Human S. E.; and Provan, K. G. (1997). *An emergent theory of structure and outcomes in small-firm strategic manufacturing networks. Academy of Management Journal*, vol. 40, no. 2, pp. 368-403, Apr.

Instituto Brasileiro de Geografia e Estatística (2004), *INDICADORES de desenvolvimento sustentável*, Brazil. Available at: <www.ibge.gov.br>. (Access: 25 Nov. 2008).

International Organization for Standarization (2008). ISO/TMB WG SR N 157. *Switzerland.*

Marchi, Jamur Johnas; Cassanego Junior, Paulo; and Gregori, Roberto de (2007). *Troca e criação de valor: possibilidades competitivas advindas da estratégia de redes*. In: ENANPAD, 31., Rio de Janeiro. *Resumo dos trabalhos*. Rio de Janeiro.

Marteleto, R. M. (2001), *Análise de redes sociais: aplicação nos estudos de transferência da informação*. Ci. Inf., Brasília, vol. 30, no. 1. Available at: <http://www.scielo.br/scielo.php?script=sci_arttext&pid=S0100-19652001000100009&lng=pt&nrm=iso>. (Accessed: 15 Feb. 2009).

Migliori, Regina de Fátima (2009). *Comunicação sustentável e redes de relacionamento*. In: Ciclo Comunicar Sustentabilidade, Rio de Janeiro. *Trabalhos*. Rio de Janeiro: ESPM.

Moore, Geoff (1999). *Tinged shareholders theory: or what's so special about stakeholders? Business Ethics: A European Review*, Osford, vol. 8, no. 2.

Perim, Juliana Andreão; and Zanquetto Filho, Hélio (2007). *Formação de rede interorganizacional para a gestão da cadeia de suprimentos: o caso do setor avícola no Estado do Espírito San Cândido*. In: ENANPAD, 31., Rio de Janeiro. *Resumo dos trabalhos*. Rio de Janeiro: [s.n.].

Petrobras (2008). *Planejamento estratégico e tático do Ceap Amazônia*. Rio de Janeiro.

Sachs, I (2002). *Caminhos para o desenvolvimento sustentável*. Rio de Janeiro: Garamond.

Stewart, T. A (1998). *Capital intelectual*. Rio de Janeiro: Campus.

Volpon, Claudia Torres; and Macedo-Soares, T. D. L. V. A. de (2007). *Alinhamento estratégico da responsabilidade socioambiental corporativa em empresas que atuam em redes de relacionamento: resultados de pesquisa na Petrobras. RAP*, Rio de Janeiro, vol. 41, no. 3, pp. 391-418, May/Jun.

2013, Impact of Global Climate Change on Earth Ecosystems *Pages* **13–20**
Editors: **D.R. Khanna, A.K. Chopra, Gagan Matta, Vikas Singh & Rakesh Bhutiani**
Published by: **BIOTECH BOOKS, NEW DELHI**

Chapter 2

Climate Change Impact on Indian Agriculture

Shashikant Rokade, Ashish Lambat and Raghoba Nagpure
Sevadal Mahila Mahavidyalaya, Nagpur, Maharashtra

In India agriculture is hardly affected due to the climate change or global warming. The success of Indian agriculture much depends on the normal monsoon as also on favorable weather conditions. The fact is that all over the world, wherever, whenever the farmers meet, the crops and the prevailing prices. Aberrant weather conditions worry them. During recent years, burning of fossil fuels by the vehicles, coal burning by the power plants, emission from the industrial factories as also as a result of the large scale deforestation, the earth's surface temperature is increasing, rainfall pattern is shifting causing loss of moisture, occurrence of cyclones, thunderstorms, floods as also the rising of the sea-level, which may ultimately see many coastal cities and towns inundated. The climate change causes a greater impact on the agriculture. As far as food grains production is concerned, even a slight rise in the earth's surface temperature could cause drop in the country's wheat production substantially as also it could much impair the quality of rice (specially of the basmati rice), fruits, vegetables and medicinal plants products, which are now being much valued for their export. In a country like India where seventy percent of the population is dependent on agriculture, it is imperative that the effect of such drastic changes in environment is studied. Also, it is equally important that we rely more on scientifically proven facts about global climate changes rather than mere conjectures and exaggerations. In this paper climate change impacts in agriculture are examined with particular to Indian scenario.

Introduction

In the country where seventy per cent of the population is dependent on agriculture, it is imperative that the effect of such drastic changes in environment is studied. Also, it is equally important that we rely more on scientifically proven facts about global climate changes rather than mere conjectures and exaggerations.

Agriculture is one of the most weather-dependent of all human activities. In India, agriculture and allied activities constitute the single largest component of Grass Domestic Product (GDP) contributing nearly 25 per cent of the total. The tremendous importance of this sector to the Indian economy can be ganged by the fact that it provides employment to to-thirds of the total workforce. The share of agricultural products in exports is also substantial, with agriculture accounting for 15 per cent of export earnings. Agricultural growth also has a direct impact on poverty eradication, and is an important factor in employment generation. Further, Indian agriculture is fundamentally dependent on weather for higher productivity.

What Is Climate Change?

Climate change can manifest itself in gradual changes in temperature, precipitation and a rise in sea level, resulting in changes in frequency, intensity and duration of extreme events. Global warming means earth getting warmer and resulting in an´ecological imbalance.

Climate Change Effect on Agriculture: An introduction

Indian agriculture is fundamentally dependent on weather for higher productivity. The proof of this has been the increasing in agricultural production, owing to good monsoons over the last few years. A few conclusions on the effect of climate change on agriculture from different studies are listed below:

- ☆ Agarwal and Sinha (1993) showed that a 2°C temperature rise would decrease wheat yields in most places.
- ☆ Rao and Shina (1994)-showed that wheat yields could decrease between 28-68 per cent without considering the CO_2 fertilization effects.
- ☆ Sinha and Swaminathan (1991) – showed that an increase of 2°C in temperature could decrease the rice yield by about 0.75 ton/ha (hectares) in the high yield areas; and 0.5°C increase in winter temperature would reduce wheat yield by 0.45 tons/ha.
- ☆ Saseendran *et al.* (2000) showed that for every one degree rise in temperature the decline in rice yield would be about 6 per cent.

Beyond a certain range of temperatures, warming tends to reduce yields because crops speed through their development, producing less grain in the process. Evaporation from the soil accelerates when temperatures rise and plants increase transpiration–that is, lose more moisture from their leaves. The combined effect is called "evapotranspiration." Because global warming is likely to increase rainfall, the net impact of higher temperatures on water availability is a race between higher evapotranspiration and higher precipitation. Typically, that race is won by higher evapotranspiration. But a key culprit in climate change–carbon emissions–can also help agriculture by enhancing photosynthesis in many important, so-called C3 crops (such as wheat, rice, and soybeans). The science, however, is far from certain on the benefits of carbon fertilization. But we do know that this phenomenon does not much help C4 crops (such as sugarcane and maize), which account for about one-fourth of all crops by value.

India, as a developing country has reasons to be concerned about the adverse impact of climate change on its economy. A large part of its population depends on climate sensitive sectors for livelihoods which makes it highly vulnerable to climate change. Climate change can have serious impact on its crops, forests, coastal regions, etc. which can in turn affect the achievement of its important national development goals. The issue of climate change cannot however be taken up without linking it to developmental needs such as poverty, health, energy access and education. Efforts to address climate change adaptation and mitigation needs should not take resources away from the core development needs and growth objectives of the developing countries. Climate Change mitigation and poverty reduction should be addressed simultaneously.

The Food and Agricultural Organization (FAO) has warned that India could lose upto 125 million tones of cereals. The Intergovernmental Panel on Climate Change (IPCC) which was awarded Nobel Prize in the year 2008 has warned that due to global warming, the sea levels in Asia will rise by atleast 40cm by 2100 flooding vast areas on the coastline, including some of the most densely populated cities, whose population will be forced to migrate inland (The Hindu Business Line dated 03-11-2007). The first to be affected by the climate change will be those living in the fringe zones, working outdoors or whose livelihoods depend directly on the physical environment. Viewed from these perspectives, the main sectors facing risks from direct impact could be agriculture, water resources and health.

The IPCC report indicates high probability of crop losses with increase in temperature in the tropical regions. Studies do confirm this trend. Among cereal crops important for food security, wheat is most sensitive to even small increase in temperature. Relatively, rice has greater tolerance to temperature increases. Increasing climate variability could result in considerable season/ annual fluctuation in food production. All agricultural commodities are subject to such variability.

The above facts emphasize the need to not only study in detail the climate change vulnerability of agriculture but also the methods of improving the adaptive capacity of agriculture to climate variability and extremes.

The Green House Effect

Under normal circumstances the sun's rays hit the earth and are reflected back into space. However, gases in the atmosphere such as carbon dioxide and methane form a barrier for sun light. Because of this property of these gases the reflected rays of the sun are trapped in the atmosphere. The sun rays cannot escape from the earth's atmosphere, and the earth heats up. This is called the green house effect.

Global Warming Effect

Global warming is increasing the earth's average temperature. The green house gases are the main culprits of the global warming. The green house gases like carbon dioxide, methane and nitrous oxide are playing hazards in the present time. These green house gases trap heat in the earth's atmosphere and thus result in increasing the temperature of the earth. The excessive emission of these gases is the major cause of global warming.

The global warming has led to increase in mean earth surface temperature and thus melting of the polar ice. There are frequent meltdown of glaciers that result in floods and other natural calamities. The melting of ice at the poles has led the increase in sea level. And further increase in temperature may melt the ice and lead to increase in mean sea level which will engulf low lying countries.

The effects of global warming can be felt on the seasons. There is shift in season cycle, as the summers are getting longer than the winters. This has affected the animals and made them to change their life style accordingly and those who failed to do so have perished or on the verge of extinction.

The global warming is also responsible for the introduction of some new diseases. The bacteria are more effective and multiply much faster in warmer temperatures compared to cold temperatures. The increase in temperature has led to increase in the microbes that cause diseases.

The major source of carbon dioxide is the power plants. These power plants emit large amounts of carbon dioxide produced from burning of fossil fuels for the purpose of electricity generations. Coal is the major fuel that is burnt in these power plants. Coal produces around 1.7 times as much carbon dioxide per unit of energy when flamed as does natural gas and 1.25 times as much as oil. Coal gives out 80 per cent more carbon per unit of energy it produces as compared to natural gas. Another major source of carbon dioxide in the atmosphere is the emission from the cars and other vehicles. About 20 per cent of carbon dioxide emitted in the atmosphere comes from burning of gasoline in the engines of the vehicles.

Buildings, both commercial and residential represent a large source of global warming pollution than cars and trucks. Building of these structures requires a lot of fuel to be burnt which emits a large amount of carbon dioxide in the atmosphere.

The second major greenhouse gas after CO_2 which causes global warming is Methane. Methane is obtained from resources such as rice paddies, Bovine Flatulence, bacteria in bogs and fossil fuel manufacture. Almost in all parts of the world, rice is grown on flooded fields. When fields are flooded, anaerobic situation built up the organic matter in the soil decays, releasing methane to the atmosphere. Nitrous oxide which is a colorless gas with sweet odors is another green house gas. The main sources of nitrous oxide include nylon and nitric acid production, cars with catalytic converters, the use of fertilizers in agriculture and burning of organic matter. A greater emission of nitrous oxides in the recent decades is leading global warming.

Climate Change In India

The vulnerability of Indian agriculture to climate change is well acknowledged. But what is not fully appreciated is the impact this will have on rain-fed (non-irrigated) agriculture, practiced mostly by small and marginal farmers who will suffer the most.

The crops that may be hit include pulses and oilseeds, among others. These are already in short supply and are consequently high-priced. Nearly 80 million hectares, out of the country's net sown area of around 143 million hectares, lack irrigation facilities and hence, rely wholly on rain water for crop growth. Over 85 per cent of the pulses and cereals, more than 75 per cent of the oilseeds and nearly 65 per cent of cotton are produced from such lands. The crop yields are quiet low.

According to A K Singh, Deputy Director-General (Natural Resource Management) of the Indian Council of Agricultural Research (ICAR), medium-term climate change predictions have projected the likely reduction in crop yields due to climate change at between 4.5 per cent and 9 per cent by 2039.

The long run predictions paint a scarier picture with the crop yields anticipated to fall by 25 per or more by 2099. This will have a detrimental effect on farmers' income and purchasing power, with obvious down-the-line repercussions. Though the rainfall records available with the Indian Meteorological Department (IMD) do not indicate any perceptible trend of change in overall annual monsoon rainfall in the country, noticeable changes have been observed within certain distinct regions.

At least three meteorological sub-divisions – Jharkhand, Chhattisgarh and Kerala have shown significant decrease in seasonal rainfall though some others have recorded an uptrend in precipitation as well. Since, rain-fed crops like coarse grains, pulses and oilseeds are grown mostly during the kharif season, these are impacted by both low as well as excess rainfall.

India is among countries most threatened by climate change with experts warning that rising temperatures will lead to more floods, heat waves, storms, rising sea levels and unpredictable farm yields. Here are the main potential effects of climate change on a country which is the world's seventh largest in area and is home to 1.1 billion people, a sixth of humanity.

How will Climate Change Effect on Agriculture?

Soil Processes

The potential for soils to support agriculture and distribution of land use will be influenced by changes in soil water balance.

- Increase in soil water deficits *i.e.* dry soils become drier, therefore increased need for irrigation but;
- Could improve soil workability in wetter regions and diminish poaching and erosion risk

Crops

The effect of increased temperature and CO_2 levels on arable crops will be broadly neutral:

- ☆ Horticultural crops are more susceptible to changing conditions than arable crops
- ☆ Field vegetables will be particularly affected by temperature changes
- ☆ Water deficits will directly affect fruit and vegetable production

Table 2.1: Predicted Effects of Climate Change on Agriculture Over the Next 50 Years

Climatic Element	*Expected Change by 2050*	*Confidence in Prediction*	*Effect on Agriculture*
CO_2	Increase from 360 ppm to 450-600 ppm	Very high	• Good for crops • Increased photosynthesis • Reduced water use
Sea level rise	Rise by 10-15 cm	Very high	• Loss of land • Coastal erosion • Flooding • Salnization of groundwater
Temperature	Rise by 1-2ºC increase frequency of heat waves	High	• Faster, shorter earlier growing seaons • Heat stress risk • Increase evapotranspiration
Precipitation	Seasonal changes by ±10 per cent	Low	• Impacts on drought • Risk soil workability • Water logging
Storminess	Increased wind speeds, especially in north. More intense rainfall events	Very low	• Lodging • Soil erosion • Reduced infiltration of rainfall

Improving Adaptive Capability of Agriculture

The following actions could be helpful in improving the adaptive capability of agriculture:

1. Improved training and general education of populations dependent on agriculture
2. Agricultural research to develop new crop varieties
3. Identification of the present vulnerabilities of agricultural systems
4. Food programs and other social security programs to provide insurance against supply changes
5. Transportation, distribution and market integration to provide the infrastructure to supply food during crop short falls.

In addition to the above improvements, it is imperative that the developed countries and the rapidly developing countries formulate strategies to curb green house gas emissions. Countries on the fast track of economic growth should also look at adopting new energy-saving technologies and planting of more trees. The emphasis should also be laid on increasing the use of renewable energy sources like solar and wind. It is high time for leading emitters of CO_2 to formulate national programs to address climate change. Only then the effect of climate change on agriculture can be reduced.

Conclusion

The policy implications for climate change impacts in agriculture are multi-disciplinary, and include possible adaptations to account for changing crop yields (increasing in some areas and decreasing in others) as well as shifting boundaries for crops, and the impact that this can have on food supply. Trade policy: changes in certain crops can affect imports/exports, depending on the crop (this is particularly relevant for cash crops such as chillies). Livelihoods: With agriculture contributing significantly to GNP, it is critical that policy addresses issues of loss of livelihood with changes in crops, as well as the need to shift some regions to new crops, and the associated skills training required. Water policy: Because impacts vary significantly according to whether crops are rain fed or irrigated, water policy will need to consider the implications for water demand of agricultural change due to climate change. Adaptive measures: Policy-makers will also need to consider adaptive measures to cope with changing agricultural patterns. Measures may include the introduction of the use of alternative crops, changes to cropping patterns, and promotion of water conservation and irrigation techniques.

Due to the complex interaction of climate impacts, combined with varying irrigation techniques, regional factors, and differences in crops, the detailed impacts of these factors need

to be investigated further. Specific recommendations for further research include: Precision in climate change prediction with higher resolution on spatial and temporal scales; Linking of predictions with agricultural production systems to suggest suitable options for sustaining agricultural production; Preparation of a database on climate change impacts on agriculture; Evaluation of the impacts of climate change in selected locations and Development of models for pest population dynamics.

Meeting the challenge of global warming will require sustained effort over decades on the part of Governments, who must establish the rules and modify them. Although much is being done to reduce the emission of green house gases, the efforts are still not enough. The Indian policy makers must remember that climate change has a creeping effect on the economy and can further hurt the already fragile agricultural system.

References

Auffhammer M, Ramanathan V, Vincent JR (2006) Proc Natl Acad Sci USA 103:19668–19672.

Chameides WL, Yu H, Liu SC, Bergin M, Zhou X, Mearns L, Wang G, Kiang CS, Saylor RD, Luo C, *et al.* (1999) Proc Natl Acad Sci USA 96:13626–13633.

Krishna Kumar K, Rupa Kumar K, Ashrit RG, Deshpande NR, Hansen JW (2004) Int J Climatol 24:1375–1393. CrossRef

Lal M, Harasawa H, Murdiyarso D, Adger WN, Adhikary S, Ando M, Anokhin Y Cruz RV, Ilyas M, Kopaliani Z, *et al.*, McCarthy JJ, Canziani OF, Leary NA, Dokken DJ, White KS (2001) in Climate Change 2001: Impacts, Adaptation, and Vulnerability, eds McCarthy JJ, Canziani OF, Leary NA, Dokken DJ, White KS (Cambridge Univ Press, Cambridge, UK), pp 533–590.

Mall RK, Singh R, Gupta A, Srinivasan G, Rathore LS (2006) Clim Change 78:445–478. CrossRef

Peng S, Huang J, Sheehy JE, Laza RC, Visperas RM, Zhong X, Centeno GS, Khush GS, Cassman KG (2004) Proc Natl Acad Sci USA 101:9971–9975.

Ramanathan V, Chung C, Kim D, Bettge T, Buja L, Kiehl JT, Washington WM, Fu Q, Sikka DR, Wild M (2005) Proc Natl Acad Sci USA 102:5326–5333.

2013, Impact of Global Climate Change on Earth Ecosystems *Pages 21–30*
Editors: **D.R. Khanna, A.K. Chopra, Gagan Matta, Vikas Singh & Rakesh Bhutiani**
Published by: **BIOTECH BOOKS, NEW DELHI**

Chpater 3

Conservation of *Aconitum heterophyllum*: The Critically Endangered Species of Indian Himalayas–Status and Future Needs

Priyanka Siwach and Priyanka Solanki
Department of Biotechnology,
Ch Devi Lal University, Sirsa, Haryana

Aconitum species belonging to family Ranunculaceae are found in the Great Himalayas spread over India, Pakistan, Nepal, Bhutan, and South Tibet, where they are used in indigenous and traditional systems of medicine. The aconites in the Himalayas are usually found in the wet alpine zone, however, in Kashmir Himalayas these have been reported to occur in the temperate zones. There are about 300 species of *Aconitum* in the world. Out of these, *Aconitum heterophyllum* holds very important position. It is one of the most used non-poisonous aconites and distributed within the country in Jammu and Kashmir, Himachal Pradesh, and Uttarakhand. The tuberous roots are used in Ayurvedic and Unani formulations of medicine. *A. heterophyllum* has become endangered in the region of U.P. Himalayas (Uttarakhand Himalayas). With respect to threat and trade information, the species has been given the critically endangered (CR) status by the IUCN (..). Various efforts, *in situ* as well as *ex-situ*, are being taken for conserving this valuable plant species. The present efforts are being faced by some challenges like illegal harvesting from wild, poor response of farmers towards cultivation, improper market conditions etc. Further there is also a lack of proper implementation of biotechnological approaches for mass propagation and secondary metabolite extraction. Some steps are being recommended in the end of the paper to further strengthen the on-going conservation efforts for *Aconitum heterophyllum*.

Keywords: *Critically endangered, Medicinal plant, In-situ conservation, Ex-situ conservation, Atis.*

Introduction

India is one of the world's twelve mega biodiversity centers, having nearly 45000 different plant species (Bapat *et al.*, 2008). India's diversity is unmatched due to the presence of sixteen different agro-climatic zones, ten vegetation zones and fifteen biotic provinces (Samy and Gopalakrishnakone, 2007). Approximately 10,000 plant species are used for their medicinal values by traditional communities. Traditional systems of medicine still continue to be widely practiced, particularly in developing countries, on many accounts. Population rise, inadequate supply of drugs, prohibitive cost of treatments, side effects of several allopathic drugs and development of resistance to currently used drugs for infectious diseases have led to increased emphasis on the use of plant materials as a source of medicines for a wide variety of human ailments.

Ayurveda, the indigenous system of herbal medicine of India, mentions nearly 8,000 herbal remedies. The *Rigveda* (5000 BC) has recorded 67 medicinal plants, *Yajurveda* 81 species, *Atharvaveda* (4500-2500 BC) 290 species, *Charak Samhita* (700 BC) and *Sushrut Samhita* (200 BC) had described properties and uses of 1100 and 1270 plant species, respectively, in compounding different drugs formulations. *The Charak Samhita*, an age-old written document on herbal therapy, reports on the production of 340 herbal drugs for curing various diseases (Prajapati *et al.*, 2003). Unfortunately, much of the ancient knowledge and many valuable plants are being lost at an alarming rate. In India, forest cover is disappearing at an annual rate of 1.5mha/yr (ref..). What is left at present is only 8 per cent as against a mandatory 33 per cent of the geographical area (ref..). Many valuable medicinal plants are under the verge of extinction. *The Red Data Book of India* has 427 entries of endangered species of which 28 are considered extinct, 124 endangered, 81 vulnerable, 100 rare and 34 insufficiently known species (Thomas, 1997).

The Genus *Aconitum*

In the long list of medicinal plant species of Red Data Book, many species of the genus *Aconitum*, commonly known as monkshood and Devil's helmet, find a place in the critically endangered/endangered/vulnerable category (Srivastava *et al.*, 2010). Aconites are reputed for their medicinal and pharmaceutical value. The genus Aconitum comprises of 400 species, few are ornamental species while majority are medicinal in nature (Utelli *et al.*, 2000).The genus Aconitum belongs to family Ranunculaceae. The plants of this genus are tall, erect, stem being crowned by racemes of large and eye catching blue, purple, white, yellow or pink zygomorphic flowers.

These are herbaceous perennial plants growing in moisture retentive but well draining soils of mountain meadows. These medicinal plant species are a rich source of diterpene alkaloids used in oriental medicine (Bisset, 1981) and flavonoids studied in last ten years as chemotaxnomic markers (Lim *et al.*, 1999), many of which exhibit broad spectrum activities. The genus aconitum is distributed in western, central and Northeastern Himalayas. Among various species of this genus, *Aconitum heterophyllum* Wall. ex Royle. (Staph 1905) is reported to have a number of diverse medicinal applications and has maximum commercial demand in the national as well as international herbal market. The over-exploitation of natural resources of this species has made it a critically endangered species of Indian Himalayas.

Aconitum heterophyllum

Trade names of *Aconitum heterophyllum* are Atees, Atish, Atibish, Atvika, Ati Vasa. This is one of the most used non-poisonous aconites and distributed within the country in Jammu and Kashmir, Himachal Pradesh, and Uttaranchal. *Aconitum heterophyllum* Wall is a high altitudinal medicinal plant, commonly found in sub-alpine and alpine zone of Himalayas from Indus to Kumaon at 2000-5000m (6000-16000 ft.)

Medicinal Value

Tubers of *Aconitum heterophyllum* are cooling in potency and bitter in taste. These are used as expectorant, frbrifuge, anthelmintic, anti-diarrhoeal, anti-emetic and anti-inflammatory. They are also used against poisoning due to scorpion or snake bite and to cure fever and contagious diseases. The aqueous extract of the root induces hypertension through action on the sympathetic nervous system and in higher doses it becomes lethal.

Varieties

Although no variety has been developed for this plant, yet based on the colour of the tubers, *Aconitum heterophyllum* is classified into white, yellow, red and black varieties. The white (daughter tuber) variety, with rapid growth and high yield, is considered to be the best.

Chemical Constituents

In tubers, atisine content varies from 0.19 per cent to 0.27 per cent. However, 0.79 per cent of total alkaloids of the plant are found in tubers. Other alkaloids found in *A. heterophyllum* are heteratisine (0.3 per cent), histisine, heterophyllisine, heterophylline, heterophyllidine, atidine, and hitidine. Aconitic acid, tannic acid, a mixture of oleic, palmitic, and stearic acids, glycerides, and vegetable mucilage is also present in addition to starch and sugars.

Threat to Natural Genetic Resources of *Aconitum heterophyllum*

Aconitine and Atisine are marker components that are isolated from *A. heterophyllum* and are commercially used. Alkaloids isolated from *A. heterophyllum* Wall have been reported to show significant antibacterial, antipyretic, enzyme inhibition activity etc (Ahmad *et al.*, 2008; Anwar *et al.*, 2003; Nisar *et al.*, 2009).

Indiscriminate and unscientific harvesting of tubers from wild plants of *A. heterophyllum* has reduced this high value plant species towards rarity and is now identified as critically endangered (CAMP, 2003; IUCN, 1993; Nautiyal *et al.*, 2002). For example, in the Johar valley in the State of Uttaranchal, collectors used to collect about 200 grams of dry Atish (*A. heterophyllum*) in one day before 1998 and during 2003-2004 this amount reduced to 70-100 grams a day (Belt *et al.*, 2003; Alam and Belt, 2004). Shah (1997 and 1998) reported that *A. heterophyllum* has become a much depleted species in the Indian Himalayas and suggested its systematic cultivation.

Conservation of *Aconitum heterophyllum* in India: Status and Challenges

The world conservation strategy (IUCN, UNEP and WWF, 1980) defines conservation as "the management of human use of the biodiversity so that it may yield the greatest sustainable benefit to present generation while maintaining its potential to meet the needs and aspirations of future generations". Government of India and of various Himalayan states is serious about conservation of threatened medicinal plants and many efforts are being done in this direction. In 1985 a committee of experts was constituted by the Forest Department, Goverment of Uttar Pradesh. The committee recommended the ban of the collection of 34 medicinal species for a period of 4 years. Among these, one was *A. heterophyllum*. This was under Govt. order no.535/1-9-20 dated Jan 1986. In year 1994, the Goverment of India, Ministry of Commerce through circular Public Notice No. 47(PN)/92-97 dated 30th March 1994, prohibited the export of 56 plant species. First species in the list was *Aconitum heterophyllum* and its derivatives and extracts. The list was further amended through Notification no. 24 (RE-98)/1997-2002, in which 29 more plant species were banned for export.

Besides prohibiting the collection and export of this species, a number of other efforts are also being carried out for conserving this species. Broadly, these different conservation strategies fall in two categories: *in situ* conservation and *ex-situ* conservation. *In situ* conservation allows evolution to continue within the area of natural occurrence, and *ex situ* conservation provides a better degree of protection to germplasm compared to *in situ*. However, both *in situ* and *ex situ* conservation are complementary and should not be viewed as alternatives (Wang *et al.*, 1993).

In situ Conservation

The ultimate aim of *in situ* conservation of a species is to maintain a broad genetic diversity so that it can retain its potential to adapt to changes in the environment. At present, about 10 million hectares *i.e.* 4.5 per cent of the geographical area of India are under the *in situ* conservation program (Singh and Chowdhery, 2002). Different Himalayan states of India have established a number of national parks and wildlife sanctuaries for the conservation of wild flora and fauna of the respective area.

Medicinal plants provide a significant source of income for the rural poor of tribel belt and Himalayan region. The common practice is to collect the plants from the wild (up to 90 per cent from wild) and sale it to the contractor or the middleman.

The poor locals of the area are heavily dependent on contractors, who, unfortunately, have only commercial notions but no responsibility towards natural genetic resources. This practice is widespread and how to break the hold of contractors over collection of medicinal plants continues to be an important challenge. Some states have come up with new programs like contract to collect the medicinal plants is given to village community organizations, called "samooh". Unfortunately, the contractors have set up their own "samoohs" and this has enabled them to maintain their control over collection activities. The collection of medicinal plants is dominantly illegal and collectors, very frequently, encroach the area of national parks and wildlife sanctuaries. Certain state governments and forest departments have started the practice of appointing "van-rakshak", from the local community, but due to economic reasons, these van-rakshaks also fall in the trap of middleman and contractors. A constant increase in the global as well as domestic demand of *A. heterophyllum* has led to depletion of its natural population. Controlling the over and illegal harvesting of medicinal plants has continued to be a major challenge for the conservation.

The other step towards in situ conservation of *A. heteroplyllum* is through cultivation, which has to be encouraged for meeting the increasing demand. It is estimated that the present demand of tubers of *A. heterophyllum* in India is more than 40 tonnes p.a. to manufacture the Ayurvedic and Unani medicines, as per the market trends of 2006-07. In nature, the species is not sufficiently available to meet the demand of manufacturers, and the species has not yet been brought under cultivation anywhere in the Himalayas on a commercial scale. Cultivation through seed in *A. heterophyllum* is difficult (http://mpc.frlht.org.in/newsletter.htm) due to poor seed availability, seed dormancy and lack of superior germplasm (Nautiyal *et al.*, 2009). Limited data and literature is available on the study of seed dormancy of *A. heterophyllum*. Germination is said to be complete when the structure called radicle penetrates the area surrounding the embryo, rest of the events including mobilization of major storage reserves are associated with the growth of seedling (Bewley, 1997). But some seeds have underdeveloped embryo, which has to grow to a critical length before germination can occur. Failing to grow may lead to dormant seed. A thorough research therefore is necessary for improvement in the seed germination and dormancy breaking system. Seed germination in *A. heterophullum* is restricted by different kinds of environmental factors or due to underdeveloped embryo which lead to morphological or physiological dormancy. Hot water treatment has shown reasonable effect on the germination of *A. heterophyllum* Wall (Pandey *et al.*, 2000; Pandey *et al.*, 2005). Chilling has also found to be the crucial factor for germination of *A. heterophyllum* Wall (Beigh *et al.*, 2005).

National Medicinal Plant Board (NMPB), the apex body for medicinal plant sector in India, has sanctioned 4254 projects under two major schemes *viz.* promotional and contractual farming from April 2001 to 2006. Under this scheme 19 projects which make 2.92 per cent of total project scheme were assigned for *A. heterophyllum*, covering area of marketing, post-harvest processing and tissue culture (Kala and Sajwan, 2007). *A. heterophyllum* cultivation is also promoted by the state governments of Arunachal Pradesh, Himachal Pradesh, Meghalaya and Uttarakhand. (CUTS: Consumer Unity and Trust Society, Centre for International Trade, Economics and Environment, India.). *A. heterophyllum* is one of the 34 medicinal crop plants which are prioritized for cultivation by the National Medicinal Plant Board (Kala and Sajwan, 2007). The cost- benefit involved in cultivation of *A. heterophyllum* is presented in Table 3.1. The area under cultivation of *A. heterophyllum* is showing an increasing trend (2006 onwards) - 13 acres (2002-03), 55 acres (2003-04), 1014 acres (2004-05), 215 acres (2005-06) (Kala and Sajwan, 2007).

Table 3.1: The Factors Related to Seed Performance, Cost Involved and the Income Generated During Cultivation of *Aconitum heterophyllum*

Seedling survival (per cent)	60-70
Seed required per hectare (gm)	614
Total cost of cultivation (US $)	2258
Total production (kg/hectare)	83
Income (US $)	4528
Net income (US $)	2270

Since it is one of the most demanded plant species in the herbal market, the cultivation of this species needs more encouragement and support.

Ex-situ Conservation

It has gained international recognition with its inclusion in Article 9 of the Convention on Biological Diversity (Glowka *et al.*, 1994). *Ex situ* conservation includes herbal gardens, germplasm banks, DNA banks and techniques involving tissue culture, cryopreservation. Herbal gardens/Vanaspati Vans, established at alpine and sub-alpine zones, offer a viable method for conservation of this species (Table 3.2). Further government of Himachal Pradesh state has declared one village, adjoining to the Dhumreda Harbal Garden (Rohru), as a Atis village for promotion of ex-situ cultivation of this species due to its high potential and natural habitat. However, herbal gardens are vulnerable to natural calamities, extreme weather conditions and sudden outburst of forest fires.

Biotechnology has offered many invaluable tools for the conservation of plant genetic resources. *In vitro* propagation or micropropagation is a viable alternative for species which are difficult to regenerate by conventional methods; where population have decreased due to overexploitation by destructive harvesting and can effectively be used to meet the growing demand for clonally uniform elite plants. The *in vitro* propagation of *A heterophyllum* has been carried out by certain groups. Shoot regeneration and elongation was observed by culturing nodal segments of *A. hetrophyllum* on Murashige and Skoog (MS) medium supplemented with 0.25 mg/l BAP (Giri *et al.*, 1993). Callus induction from explants of *A. heterophyllum* was reported at extremely low concentration of growth regulators (0.5 mg/l NAA and 0.25 mg/l BAP) supplemented to MS medium (Jabeen *et al.*, 2006). Production of secondary metabolites through genetic engineering of plant in vitro is another way to cope up with the increasing demand of medicinal plants. A protocol of hairy root production in *A. heterophyllum* Wall using *Agrobacterium rhizognes* has also been standardized

(Giri *et al.*, 1997) and it was reported that total alkaloidal (aconites) content of transformed roots was 3.75 per cent times higher than non transformed roots.

Table 3.2: Some Important Herbal Gardens in the Himalayan States of India

Name	*Linked to*	*State*
State Biodiversity Park Herbal garden	State Biodiversity Park	Sikkim
Samdruptse Harbal garden	Tendong Nature reserve	Sikkim
Maenam Haerbal garden	Maenam Sanctuary	Sikkim
Ratey Chu Herbal garden	Rate Chu Reserve Forest	Sikkim
Botanical garden	Dr V S Parmar University of Horticulture and Forestry, Nauni-Solan	Himachal Pradesh
Herbal Garden-Trikuta Hills	School of Biotechnology, Shri Mata Vaishno Devi University, Kakryal, Katra	Jammu and Kashmir
Herbal garden	Research Institute in ISM, Joginder Nagar, Mandi. Research Institute in ISM, Hamirpur	Himachal Pradesh
Medicinal Plant Garden R.R.I	RRI (Ayurveda), Barsojai, Beltola, Guwahati	Assam
Medicinal Plant Garden cum Gene Pool conservation Plot	– Sangmein, Upper Shilong, Sangmein. – Umsaw Nongladew, Silviculture range	Meghalaya
Medicinal Plant Garden	Ayurveda RRI, Itanagar	Arunachal Pradesh
North-East Ecological Park	North east Institute of Science and Technology, Jorhat	Assam
Dhumreda, Rohru	Shimla	Himachal Pradesh

Unfortunately all the *in vitro* propagation work has remained limited to the laboratory level. No large scale commercial applications have been designed from these studies.

The Seed Gene Bank at Institute of Integrative Medicine (IIIM), Jammu is having entries of *A. heterophyllum* seeds. Since the seeds are dormant in nature having low germination rate, there is a need to carry out the cryopreservation of vegetative propagules of it. To the best of our literature survey, we could not find a single report on the cryopreservation of vegetative parts of the species. More efforts are needed towards applications of biotechnological techniques for the conservation of this greatly demanded medicinal crop of India.

Future Aspects

More support and encouragement is needed for popularization of *A. heterophyllum* cultivation. Some more research is needed to understand the its seed biology to enhance the frequency of seed germination and seedling survival. This is necessary to enhance the systematic cultivation of A heterophyllum. Efforts are also required to suggest appropriate cropping patterns for the incorporation of *A. heterophyllum* into the conventional agricultural and forestry cropping systems. The resources devoted to the development and diffusion of cultivation technologies need to be augmented. Similarly, efforts to produce planting material need to be increased. Further information regarding license acquisition for cultivation of medicinal plants should be well disseminated to avoid the exploitation of farmers by middleman. It is important that innovative marketing mechanisms are instituted so that farmers' risk is reduced and their income is increased. The role of NGOs in the marketing of medicinal plants should be expanded. Furthermore, marketing by farmers' associations and cooperatives need to be promoted. In order to reduce farmers' risk, insurance schemes should be introduced.

Various *In situ* conservation practices can be accomplished more efficiently by the active support of people of nearby communities in the area. Herbal nursery should be established in every Gram Panchayat Unit. The Herbal Plant Conservation Group at local level can tie up with the ayurvedic companies in the nearby cities. Under this, group training to farmers should be given to make them aware of the use of herbal plants. Self help groups formed by women should be involved for the promotion of herbal drugs from the kitchen stock and rare medicinal plants. There should be training on various aspects of medicinal plants *viz. identification, utility, application, cultivation/collection and marketing*. Organizing workshops, field's trips and practice and demonstration to NGOs, Mahila Mandals, Health Workers, Forest People, Ayurvedic students, Ayurvedic doctors, Pharmaceutical and general people on medicinal plants.

The policies pertaining to the collection of medicinal plants, and their implementation, should be made more transparent. Efforts should be done to minimize the role of middleman and contractors.

The in vitro propagation of this species should be promoted. The large scale commercial production of plantlets of this species should be initiated at identified centres. Further characterization and isolation of active principle of a. heterophyllum from callus cultures and hairy roots should be initiated for further commercial applications.

Though a number of effective steps have been taken to conservae this particular species, we suggest and conclude more attention to be paid for preserving this valuable plant for tha generation to come.

References

Ahmad M W, Ahmad M, Ahmad M, Zeeshan O and F. Shaheen, 2008. Norditerpenoid alkaloids from the roots of *Aconitum heterophyllum* Wall. with antibacterial activity. J Enzyme Inhibition Med Chem, 23: 1018-1022.

Alam G and Belt J (eds.), 2004. Searching Synergy: stakeholder views on developing a sustainable medicinal plant chain in Uttaranchal, India. pp.102, KIT Publishers, Amsterdam (KIT Bulletin 359).

Anwar S, Ahmad B, Subhan M, Gul W and Nazar-ul-Islam, 2003. Biological and pharmacological properties of *Aconitum chasmanthum*. J Biol Sci, 3: 989-993.

Bapat V A, Yadav, S R and Dixit G B, 2008. Rescue of endangered plants through biotechnological applications. Natl Acad Sci Lett, 31: 201-210.

Beigh S Y, Nawchoo I A and Iqbal M, 2005. Cultivation and conservation of *Aconitum heterophyllum*: A Critically Endangered Medicinal Herb of the Northwest Himalayas. J. Herbs Spices Med Plants., 1: 47-56.

Belt J, Lengkeek A and Van-der Zant J, 2003. Cultivating a Healthy Enterprise: developing a sustainable medicinal plant chain in Uttaranchal, India. pp.56, KIT Publishers, Amsterdam (KIT Bulletin 350).

Bewley JD, 1997. Seed germination and dormancy. Plant Cell, 9: 1 055-1066.

Bisset N G, 1981. Arrow poisons in China- Part II- *Aconitum* –botany,chemistry and pharmacology. J Ethnopharmacol, 4: 247-336.

CAMP, 2003. CAMP workshop. 22-25th May, Shimla, Himachal Pradesh.

Giri A, Banerjee P S, Ahuja P S and Giri C C, 1997. Production of hairy roots in *Aconitum heterophyllum* Wall using Agrobacterium rhizogenes. In Vitro Cell Dev Biol Plant, 33: 280-284.

Giri A, Paramir S A and Kumar A P V, 1993. Somatic embryogenesis and plant regeneration from callus cultures of *Aconitum heterophyllum* Wall. Plant Cell Tiss Org, 32: 213-218.

IUCN, 1993. Draft IUCN Red List Categories. IUCN, Gland, Switzerland.

IUCN, UNEP and WWF, 1980. World Conservation Strategy: Living Resource Conservation for Sustainable Development. IUCN, England, Switzerland.

Jabeen N, Shawl A S, Dar G H and Sultan P, 2006. Callus Induction and Organogenesis from Explants of *Aconitum heterophyllum*-Medicinal Plant. Biotechnology 5 (3): 287-291.

Kala C P and Sajwan B S, 2007. Revitalizing Indian systems of herbal medicine by the National Medicinal Plants Board through institutional networking and capacity building. Current Science,vol 93, No 6.pp

Lim C E, Park J K and Park C W, 1999. Flavonoid variation of the *Aconitum jaluense* complex (Ranunculaceae) in Korea. Plant Syst Evol, 218: 125-131.

Nautiyal B P, Nautiyal M C, Khanduri V P and Rawat N, 2009. Floral biology of *Aconitum heterophyllum* Wall: A critically endangered alpine medicinal plant of Himalaya, India.Turk J Bot, 33:13-20.

Nautiyal B P, Prakash V, Bahuguna R, Maithani U C, Bisht H and Nautiyal M C, 2002. Population study for monitoring the status of rarity of three Aconite species in Garhwal Himalaya. Trop Ecol, 43: 297-303.

Nisar M, Ahmad M, Wadood N, Lodhi M A, Shaheen F and Choudhary M I, 2009. New diterpenoid alkaloids from *Aconitum heterophyllum* Wall: Selective butyrylcholinesterase inhibitors. J. Enzyme Inhibition Med Chem, 24: 47-51.

Pandey H, Nandi S K, Nadeem M and Palni L M S, 2000. Chemical stimulation of seed germination in *Aconitum heterophyllum* Wall. and *A. balfourii* Stapf: Important Himalayan species of medicinal value. Seed Sci Technol, 28: 39-48.

Pandey S, Khushwaha R, Prakash O, Bhattacharya A and Ahuja P S, 2005. Ex-situ conservation of Aconitum heterophyllum Wall, an endangered medicinal plant of the Himalaya through mass propagation and its effect on growth and alkaloid content. Plant Genet. Resour 3: 127-135.

Prajapati N D, Purohit S S, Sharma A K and Kumar T, 2003. A Handbook of Medicinal Plants, Agrobios, Jodhpur.

Samy R P and Gopalakrishnakone P, 2007. Current status of herbal drugs and the future perspectives. hdl:10101/npre.2007.1176.1.

Shah N C, 1997. Faulty Export Policy of Herbs and Crude Drugs in India. Medicinal Plant Conservation, 4: 4-5.

Shah N C, 1998. The Status of Medicinal plants in the Indian Himalayas. Workshop on "Himalayan Medicinal Plants" held at G.B. Pant Institute of Himalayan Environment and Development, Katarmal, Kosi (Almora).

Singh N P and Chowdhery H J, 2002. Biodiversity conservation in India. In: Das A. P. (ed), Perspective of Plant Diversity. Department of Boatny, North Bengal university, West Bengal, India, pp. 501-527.9-11 November 2000, Bishen Singh Mahendera Pal Singh, Dehra Dun, India.

Srivastava N, Sharma V, Kamal B, Dobriyal A K and Jadon V S, 2010. Advancement in research on *Aconitum* sp. (Ranunculaceae) under different area - A review. Biotechnology, 9: 411-427.

Stapf O, 1905. The Aconites of India: A monograph. Ann Royal Bot Garden, Calcutta. 10 (2):116-181.

Thomas, J. 1997. Medicinal and aromatic plants research in India. In UNDP, 1997.

Utelli A B, Roy B A and Baltisberger M, 2000. Molecular and morphological analysis of European *Aconitum sp* (Ranunculaceae). Plant Syst Evol, 224: 195-212.

2013, Impact of Global Climate Change on Earth Ecosystems *Pages 31–34*
Editors: **D.R. Khanna, A.K. Chopra, Gagan Matta, Vikas Singh & Rakesh Bhutiani**
Published by: **BIOTECH BOOKS, NEW DELHI**

Chapter 4

Review of Environmental Law for E-Waste in India

Anita S. Jadhav and Sheela Warbuuwan
Department of Zoology, ICLES'M.J. College,
Vashi, Navi Mumbai – 400 703, M.S.

Electronics are changing the lives of people everywhere. They are touching every aspect of our lives and in the wake of this 21st century revolution it is sure to damage the quality of our lives and that of generations to come. E-waste contains hazardous constituents that negatively impact the environment and human health. The developing countries are facing a huge challenge in the management of electronic waste (e-waste) which are either internally generated or imported illegally as 'used' goods., India lack adequate infrastructure to manage these wastes safely, thus these wastes are buried, burnt in the open air or dumped into the surface water bodies as there is a lack of awareness among people about its treatment and serious impacts. E-waste needs to be treated as a hazardous waste which are to be tackled. Man has to deal with his legacy which emerged into environmental relating law which is complex. This paper is an attempt to focus on the law and related issue of the e-waste.

Keywords: E-waste, Present status.

Introduction

Due to ongoing scientific and technological advancement. Electronics and information technology (IT) has become the fast growing segment of Indian industry due to which there is increased growth rate of these industries in India. Due to industrialization, globalization and urbanization and with the increase in the population there is a greater pressure on the environment. According to a study carried out by MAIT-GTZ in 2007 about 3.3 hundred thousand tones of e-waste is generated annually in India and the generation of e-waste is expected to touch 4.7 hundred thousand tonnes by 2011. The rapidly growing mobile use in India will take its toll by 2020 when e-waste from discarded phones will grow 18 times from 2007 levels, And by 2020, India's e-waste from only old computers will jump 500 per cent than as compared by to 2007.

Environmental law has become a critical means of sustainable industrial growth and development. The constitutional provisions are backed by a number of laws – acts, rules, and notifications. In 1986 the EPA (Environment Protection Act), came into force which gives a holistic framework for the protection and improvement to the environment. Thereafter a large number of laws came into existence as the problems began arising.But the classification of e- waste as hazardous in Indian legislation is unclear. This paper give the review on the law related to E-waste in India.

E-Waste Related Laws in India

Since e-waste fall under the category of both 'hazardous" and "non-hazardous waste", they are covered under both the hazardous waste regulations as well as the solid waste regulations.

India needs to have strong rules and regulations in place. Over the years, the government has instituted a number of regulations for better management of hazardous waste in the country. Some of these regulations are given below: Policies, laws and regulations applicable for the management of ewaste are:

1. The National Environmental Policy 2006
2. The Environment (Protection) Act 1986
3. The Hazardous Wastes (Management and Handling) Rules 1989 as amended in 2008

The National Environmental Policy 2006 (NEP)

Policy by the Ministry of Environment and Forests that was approved by the Union Cabinet on 18 May 2006. NEP lays emphasis on:

- ✰ Encourage reuse and recycling.
- ✰ Strengthening informal sector and providing them a legal status.
- ✰ Establish system for collection and recycling of materials to recover resources.
- ✰ Environmentally safe disposal of residues.

Presently E-Wastes are Regulated Under Hazardous Wastes (M&H) Rules, 1989 as Amended in 2003 and 2008 and are Categorized as:

- ✰ Waste generated in electronic industry Schedule 1 (Rule 31).
- ✰ Schedule 2 contaminations beyond the prescribed threshold limits applicable for qualifying as hazardous waste.

- Schedule 3 Sl.No.A1180: Waste Electrical And Electronic Assemblies (export/import).
- Schedule 3 Sl.No.B1110: Electrical And Electronic Assemblies Valid For Direct

  Reuse But Not For Recycling (export/import).

The above laws have been sourced from: *Draft* The Ewaste (Management and Handling) Rules, 2010, Environmental policy-making in India report

Objective of E-Waste Regulations

- To stop dumping and encourage recycling quickly in India.
- To put in place an effective mechanism to regulate the generation, collection, storage, transport, import, export, environmentally sound recycling, treatment and disposal reducing wastes for final disposal.
- To harmonize the hazardous waste rules with the Basel Convention of 1989.

  "Guidelines for Environmentally Sound Management of E waste"published in March 2008 by the Government of India in March 2008

Effectiveness of Law

- Rule for e-waste in Indian is introduced for environmentally management and disposal of electronic waste in the country.
- The e-waste (Management and Handling) Rule of 2011 will be brought into force by May 1, 2012. by the Ministry of Environment and Forest as part of the Environment Protection Act of India.
- This rule will is applicable to producer, consumer manufacturer, sale, purchase and processing of electronic equipment or components.
-

  The rule defines of E-waste management in a manner which shall protect health and environment against any adverse effects, which results from hazardous substance present in such wastes.
- The rule mandates government departments and producers of electrical and electronic equipment to take responsibility for the collection of e-waste produced while manufacturing or after use in the end-of-life.
- It also requires affected set up collection centers to take back systems through which consumers can return their products.

Conclusion

- As there are no specific environmental laws or regulations for e-waste management till today different States under different regulatory bodies apply to various aspects of e-wastes.
- These E-waste rules will provide guidelines that are currently being developed for further legal framework.

- ✩ The existence of an adequate legal framework will enable us to establishment a sound E-waste management system for India.
- ✩ We need to have a dedicated e-waste law in India that can stringently deal with the menace of E-waste in India.

References

Ashish Chaturvedi, Rachna Arora, Sharon Ahmed (2010) Policy Cycle – Evolution of E-waste Management and Handling Rules

Chaturvedi A., Arora, R., Khatter, V. and Kaur, J., 2007, E-waste Assessment in India –Specific Focus on Delhi. *MAIT-GTZ study*.

DIT (2003). Environmental management for Information Technology industry in India, Department of Information Technology, Government of India, pp.122-124.

Divan, S. and Rosencranz, A. (2001), Environmental Law and Policy in India, Cases, Materials and Statutes, New Delhi, India.

Kurian Joseph (2007) Proceedings Sardinia, Electronic Waste Management In India– Issues and Strategies Centre For Environmental Studies, Anna University, Chennai, India

Radha G. (2002). A Study of the Performance of the Indian IT Sector' at www.nautilus.org. accessed on 21st June 2005.

2013, Impact of Global Climate Change on Earth Ecosystems *Pages* **35–37**
Editors: **D.R. Khanna, A.K. Chopra, Gagan Matta, Vikas Singh & Rakesh Bhutiani**
Published by: **BIOTECH BOOKS, NEW DELHI**

Chapter 5

Present Status of Macro Benthos Biodiversity in Vashi Creek

Usha Anil Kumar and Anita S. Jadhav
ICLES'M.J. College,
Vashi, Navi Mumbai – 400 703, M.S.

Vashi creek is a long, deep, tidal canal in Navi Mumbai.The creek stretches almo st 15.43 km along the latitude 19°5′ 20.28°N Longitude 73°0′ 16.41° E.Since the last decade Vashi creek has witnessed enormous developmental changes,which has brought about adverse effect on the biota of the creek. The creek contains numerous sources of organic pollution including industrial sewage effluent. A survey of the creek was performed assessing the macrobenthic community for the period of Jan 2011 to Aug 2011. Macro benthos along Vashi creek reported were polycheate, a biotic index of pollution. Decapods, Crustaceans, Bivalve, Gobid Gastropod and other dominant species were also found.

Keywords: Macrobenthos, Creek, Biodiversity.

Introduction

In an aquatic ecosystem, the physico-chemical environment controls the diversity, biomass and distribution of biotic communities. Further different species of flora and fauna exhibit variations in their response to any alteration in the environment and have an indicator value.Macro benthos being non-migrant inhabitants can be used as indicators of ecological changes in the coastal water.Most important feature of the animal communities is their diversity, which explains the organisation of a community. In India, studies have been conducted on the benthic macro of different aquatic systems.(Michael,1968; Mandal and Moitra, 1975; Vasisht and Bandal,1979; Sarkar,

1989) and much of the literature available gives the biodiversity of lakes macrofauna (Krishnamoorthi 1966, 1979, Gupta 1976, Patnaiak 1971, Raman *et al.*, 1975). Thus the present attempt is made to give the data of benthic macrofaunal biodiversity along the creek.

Material and Methods

Sampling was done from five site with each area of about 2.5 m^2 was covered while recording the data. The creek was visited twice a month during the full moon day from Jan 2011 to Aug 2011. The sampling was done by using quadrant methodand shell were collected as samples for photographs.

Observation

This ecosystem harbors variety of organisms of marine, brackish water and terrestrial forms. Creek is muddy and rocky the most common macrofauna along the creek consisted of Molluscans (20) Arthropods (5), Polycheates (2), Plathylelminthes (1).Settlement of Molluscans were maximum dominated by gastropods- Potamidescingulatus, Telescopium telescopium few forms of Conus,Cerithium, Nerita oryzarium, Turitella, Thias and Oliva. Bivalvian biodiversity was dominated by Arcagranulosa, Cardiumflavum, Cardiumasiaticum, Katelysia, Crassostrea cucullata, Gaffrarium divaricate and few Meretix meretix, Cardiata, Arcabistriagata, Placuna and Mytilus. Athropods were represented by Barnacle, Scylla, Fiddler crab and Eupagarus crab. Hetroneris and Sabella were the only polycheate observed. Platyhelminthes representative observed was conuluta, The common benthic fishes observed were penaeid and non-penaeid, cat-fish and Gobid.

Conclusion

Macrobenthos biodiversity along the Vashi creek is dominated by molluscans species near the shore line. Although fiddler crabs were also abundant. If human interference and beautification drive can be minimized, the biodiversity can be conserved to sustain mangrooves which supports life.

References

Gupta, S. D. (1976). Macrobenthic fauna of Loni reservoir. J. Ind. Fish. Soc. India. 8: pp 49-59.

Krishnamoorthy, K. N. (1966). Preliminary studies on the bottom macrofauna of the Thungabhadra reservoir. Proc. Ind. Acad. Sci., B 65: pp 96-103.

Krishnamoorthy, K. P.; and Sarkar, R. (1979). Macro-invertibrates as indicators of water quality. In: Proceeddings Symposium of Environmental biology, Verma, S. R.; Tyagi, A. K.; and Bansal, S. K. (Eds.) The Academy of environmental biology, Myzaffarnagar, pp: 133-138.

Mandal, B. K.; and Moitra, S. K. (1975). Studies on the bottom fauna of a fresh water fish pond at Burdwan. J. Inland Fish. Soc. India. VII: pp 43-48.

Micheal, R. G. (1968). Studies on the bottom fauuna in a tropical fresh water fish pond. Hydrobiologia, 31: pp 203-230.

Patnaik, S. (1971). Seasonal abundance and distribution of bottom fuana of the Chilka lake. J. Mar. Biol. Ass. India 13(1): pp 106-125.

Raman, K.; K. V. Ramakrishna, S. Radhakrishnan and G. R. M. Rao. (1975). Studies on the hydrobiology and benthic ecology of lake pulicat. Bull. Dept. Mar. Sci. Univ. Cochin 7(4): pp 855-884.

Sarkar, S. K. (1989). Seasonal abundance of benthic macrofauna in a freshwater pond. Envi. and Ecol., 7(1): pp 113-116.

Vasisht, H. S.; and Bhandal, R. S. (1979). Seasonal variation of benthic fauna in some North Indian lakes and Ponds. India. J. Eco., 6(2): pp 33-37.

2013, Impact of Global Climate Change on Earth Ecosystems *Pages 39–47*
Editors: **D.R. Khanna, A.K. Chopra, Gagan Matta, Vikas Singh & Rakesh Bhutiani**
Published by: **BIOTECH BOOKS, NEW DELHI**

Chapter 6

Histopathological Effect of Testosterone Propionate on Male and Hydroxyl Progesterone Caproate on Female Rain Quail, *Coturnix coromandelica*

***Pravin Charde*[1], *Jyoti Ramteke*[1] *and Suresh Zade*[2]**
[1]*Principal, Sevadal Mahila Mahavidyalaya, Nagpur – 440 009, M.S.*
[2]*Professor, P.G.T.D. Zoology,*
Rashtrasant Tukadoji Maharaj Nagpur University, Nagpur, M.S.

Adult rain quail are sexually diamorphic. The aim of this study was to determine the effect of testosterone propionate and hydroxyl progesterone caproate on gonads and some other endocrine glands of male and female rain quail. 15 male and 15 female rain quails were treated with testosterone and progesterone daily for 7 and 15 days respectively. The study reveals that after the treatment of testosterone, the spermatogonia were detached and show hypertrophy as well as the germ cells were enlarged and vacuolated every degenerates, thyroid and adrenal shows hypertrophy.

Keywords: *Testosterone propionate, Hydroxyl progesterone caproate, Sexual dimorphism, Gonads, Rain quail.*

Introduction

Quails are almost tailless patridge like bird popularly known as "Batter", belonging to the class Aves and family *Phasianidae*. Batter is a good table bird known for its delicacy since olden days. They are used as food before chicken was domesticated. Meat is rich in vitamins, amino acids, unsaturated fatty acid etc., which are very vital for health of human being. Quail meat therefore can be recommended to be included in diet of children, pregnant mother and convalescent patient for speedy recovery.

Perusal of literature on quail reproduction reveals that, little is known about these aspects in exotic and tropical species. The reason is that more critical experiments have not been successfully undertaken in Indian Quail species. Studies on these species are confined mainly to the changes in the reproductive organs and rarely endocrine glands during different phases of reproductive cycle. No experimental studies have been made in these species endocrine glands in reproduction. Fundamental knowledge of reproductive process and the endocrine mechanism that control process is therefore, invisible for profitable management and production of these birds in large scale. The paucity of information on reproduction endocrinology of Indian quail motivated me to undertake this investigation.

The present study was undertaken on Rain Quail, *Coturnix coromandelica*. This bird was selected for this work because it offers the advantages of endocrinological works. The aim and objective this investigation is to study the histopathological effect of testosterone propionate and hydroxy progesterone caproate on gonads and other endocrine glands and to investigate the mechanism of action of this steroids in both sexes of this Rain Quail.

In the present study testosterone propionate and hydroxyl progesterone caproate obtain from German Remedies Limited, Bombay under trademark of Schering AG, West-German.

Material and Methods

The Indian Rain Quail C. *coromandelica* has been selected for present investigation. Adult quail are sexually dimorphic. Reproductive behaviour is sexually differentiated in quail. The adult rain quail is about 6½ to 7 inches in size (without tail) and body weight is 85 to 90 gms. Temperature, light, floor space, humidity and feed are most important factors for good housing and maintenance of quails. The maximum daily amount of feed given was equivalent to 20 to 25 per cent of body weight. To identify individual quails from each groups, they were marked with numbered aluminium strips using bands.

The Rain Quail utilized in this study were collected from various places in Nagpur district and Amravati district of Maharashtra State. The birds were trap over a period of four months during October to January and consisted of 60 adult males and 40 adult females weighing 70 to 90 gms. Before commencement of the experiments all birds were acclimatized in laboratory for at least 2 to 3 weeks.

The males are divided into four groups of 15 animals each. The ground II and IV are given intramuscular injection of testosterone propionate 0.15 mg/kg/day for 7 and 15 consecutive days respectively. The I and III group receive equivalent volume of physiological saline for the same period. For another experiment 40 female birds were divided into four groups of 10 birds each. The group II and IV were given intramuscular injection of hydroxy progesterone caproate 0.15 mg/kg/day consecutive for 7 and 15 days respectively. The I group and II were used as a control and treated with equivalent volume of physiological saline for the same period. Birds of experimental and control groups were weighed before and at the end of the experiments and maintained under same husbandry condition. At the end of each experimental animals were sacrificed. For histopathological studies reproductive organs and endocrine glands were dissected, weighed and fixed in various fixatives. All endocrine glands were fixed in formal sublimate for 24 Hrs. and other tissue were fixed in Bouin's fluid. The tissue were dehydrated in various grades of alcohol cleared in xylene and after embedded in paraffin blocks were prepared. Paraffin sections were cut at 5 to 6 µm thickness. Sections were stained with haematoxyline and eosin.

Histomorphology

Testis (Figure 6.1)

The testis consists of compact convoluted seminiferous tubules which are lined by single or double layer of germinal cells. The tubules are composed of primary spermatogonia, secondary spermatogonia and sectoli cells. Further stages of spermatogenesis such as spermatocytes, spermatids and spermatozoa are not observed. Similarly, the sertoli cells are inconspicuous and regressed. Since, the control males are sacrificed in the month of January which is sexually inactive period in rain quails, the testis have showed regression. In the interfollicular region blood capillaries, connective tissue and regressed Leydig's cells are present.

Ovary (Figure 6.2)

In the rain quail single ovary and oviduct is present which is situated on the left side of the cephalic end of kidneys and attached the body wall by ligament. In rain quail, the ovary contains several primary follicles and blood haemorrhage. Developing follicles are composed of germinal vesicle granulose cells, theca interna and theca externa. Some degree of atresia is observed at primary follicular stage. Rarely degeneration is visible in large follicles.

Adrenal gland of rain quail is paired, triangular in shape and yellow or orange in colour. It locates anterior and medial to the cephalic lobe of kidney and posterior to the lungs. It is surrounded by thick capsule which is made up of connective tissue.

Pineal gland is present on the dorsal side of the brain in triangular space between cerebellum and cerebral hemisphere. Pineal grain in rain quail is creamisa brown, club-shaped conical structure. Pineal gland is composed of lobules and each lobule consists of follicles which are separated by intralobular septa.

Thyroid gland of the male and female rain quail is composed of two separate lobes which are reddish in colour, situated on either side of the trachea in between

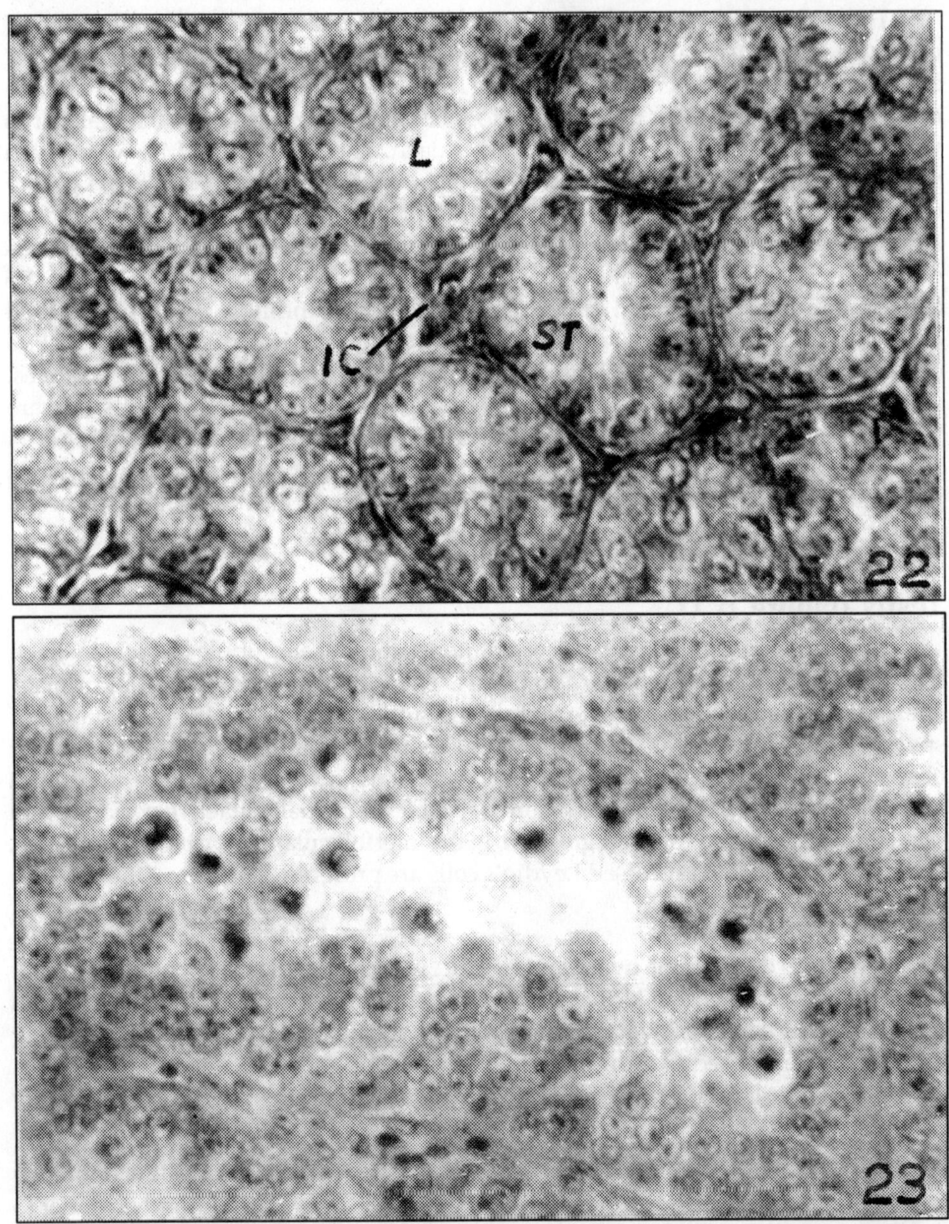

Figure 6.1: Photomicrograph of Control Testis Showing Convoluted Seminiferous Tubules with Germ Cells, Sertoli Cells and Interstitial Cells

jugular and carotid blood vessels. The normal thyroid gland is encapsulated and composed of well organized compact follicles.

Parathyroid gland in the rain quail is paired, well vascularised and closely attached to the posterior part of the thyroid gland.

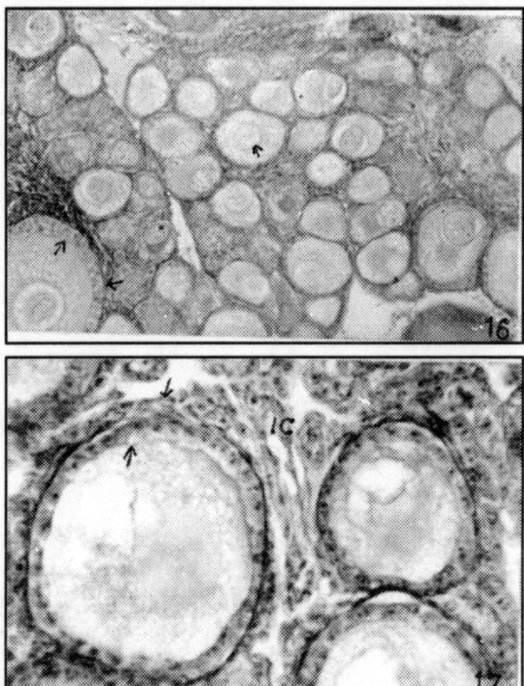

Figure 6.2: Photomicrograph of Control Ovary Showing Primary Follicles and Germinal Follicles with Theca Externa, Theca Interna and Interstitial Cells

Results and Discussion

Pineal

7 Days After Treatment

No significant changes were observed in the vasculature of pineal gland. Lumen of the follicle was slightly obliterated. Type-I *i.e.* polygonal cells shows regression whereas Type-II *i.e.* tall cells shows hypertrophy and degranulation.

15 Days After Treatment

Showed more pronounced hypertrophy of Type-II *i.e.* tall columnar cells as compared to 7 days of testosterone propionate. The cell debris in lumen was evident. Type-I *i.e.* polygonal cells showed more regression.

Thyroids

7 Days After Treatment

Testosterone propionate treatment resulted in hypertrophy of the follicular epithelial cells and thickness of follicular wall. It induced hypertrophy and hyperplasia in small follicles. Large follicles showed little regressive effect.

15 Days After Treatment

Induced changes in thyroid follicles. Diameter of follicles were increased. The gland contained several empty follicles. The colloid contained few vacuoles at the peripheral region. The reduction in the amount of colloid was more evident.

Parathyroid

7 Days After Treatment

Testosterone propionate treatment induced hypertrophy and vacuolization in the cells of parathyroid cells of parathyroid gland. Small circular vacuoles appeared adjacent to the nucleus.

15 Days After Treatment

The above mentioned effect of testosterone propionate were same, rather more pronounced were same, rather more pronounced in the fifteen days treated birds.

Adrenal

7 Days After Treatment

No significant changes were found in the vasculature of the gland. No changes were evident in Type II cells while Type I cells showed slight hypertrophy and degradation.

15 Days After Treatment

More significant changes were seen in the internal cells. The internal cells shows hypertrophy and were degranulated and vacuolated. No significant changes were observed in Type II and showed little hypertrophy as compared to control.

Testis

7 Days After Treatment

The germinal epithelium was not affected after treatment of testosterone propionate. Debris formation in the lumen of the seminiferous tubules was seen.

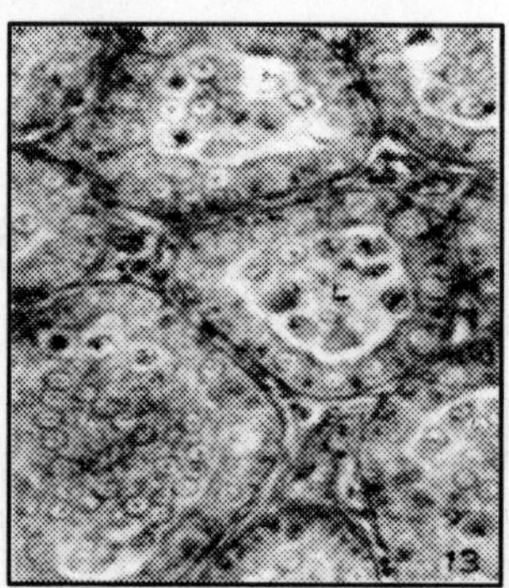

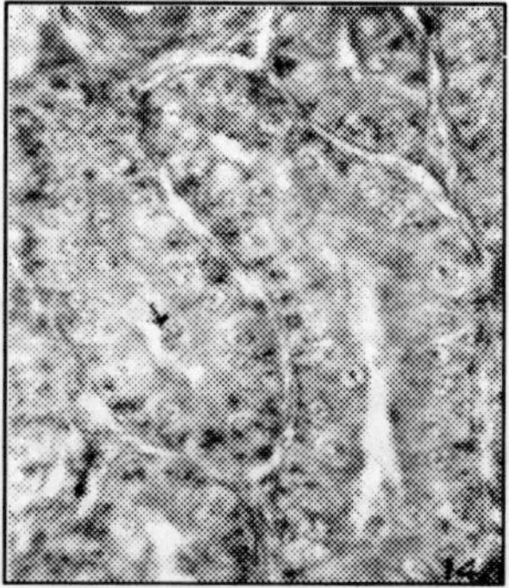

Figure 6.3: Photomicrograph of Testis 7 Days Treatment of Testosterone Propionate Showing Detached Spermatogonia in the Lumen

Regressive changes were observed in peripheral seminiferous tubules. Primary spermatogonia and secondary spermatogonia were detached and seen accumulated in the lumen of seminiferous tubules. In some tubules lumen disappeared because of hypertrophy of germ cells. No changes were observed in the vasculature of gland.

15 Days After Treatment

In most of the tubules germ cells detached and accumulated as debris in the lumen. Germ cells were enlarged and vacuolated. Lumen of seminiferous tubules

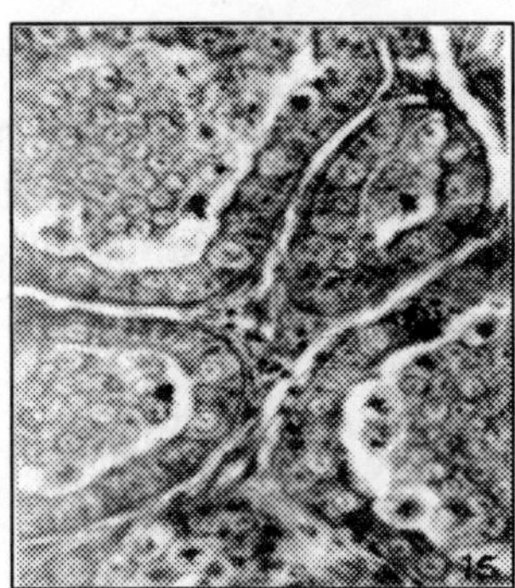

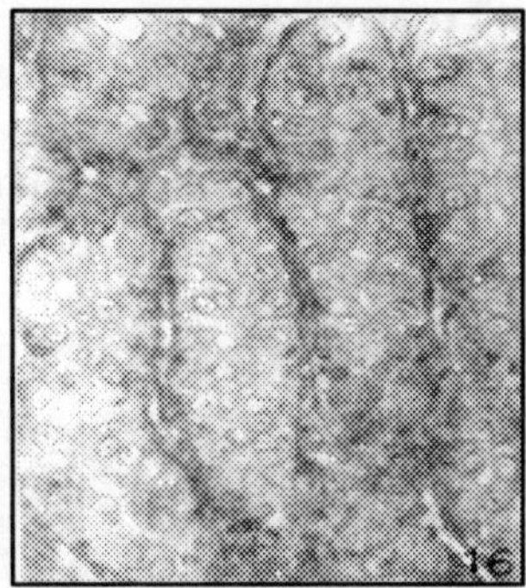

Figure 6.4: Photomicrograph of Testis 15 Days Treatment of Testosterone Propionate Showing Completely Obligated Lumen, Hypertrophy of Sertoli Cells and Degenerative Effect on Leydig's Cells (Interstitial Cells)

disappear enlarged size of the primary spermatogonia and secondary spermatogona showed hypertrophy. Reydig's cells were degenerated while sertoli cells showed hypertrophy. Blood vessels because dilated.

Ovary

7 Days After Treatment

The atretic primary follicles and formation of space between the haemorrhage and ovarian stroma changes. Haemorphage become more viscous. The large follicles shows regressive changes.

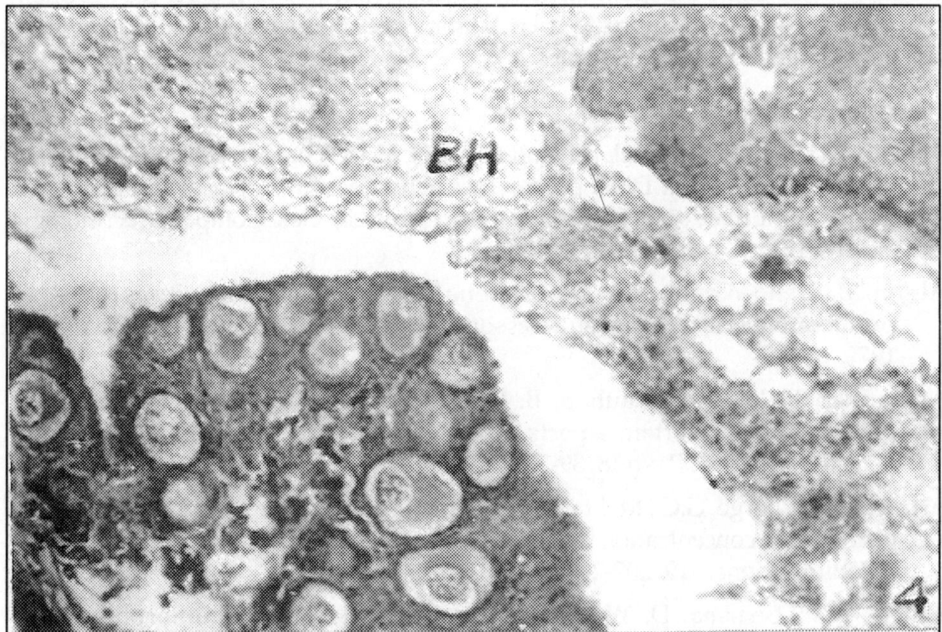

Figure 6.5: Photomicrograph of Ovary 7 Days Treatment of Hydroxyl Progesterone Caproate Showing Primary Follicles and Formation of Space Between Ovarian Stroma and Blood Haemorrhage

15 Days After Treatment

The space between blood haemorrhage and ovarian stroma is more pronounced. Follicular cytoplasm and germinal vesicle shows regressive changes. There was no significant effect on the theca interna and theca externa, but these two layers started to separate from each other in large follicles. In stroma interstitial cells shows more degenerative changes.

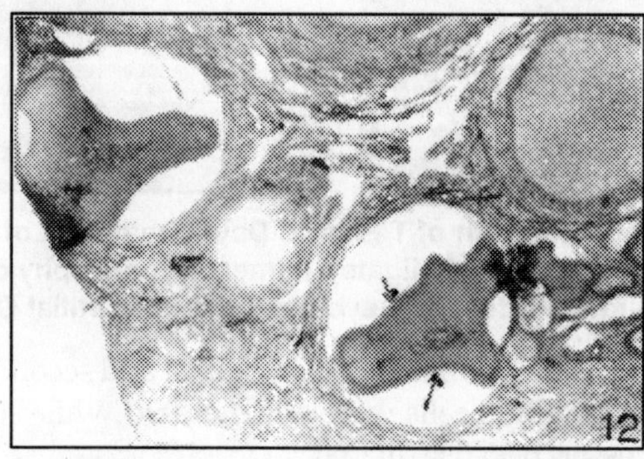

Figure 6.6: Photomicrograph of Ovary 15 Days Treatment of Hydroxyl Progesterone Caproate Showing Greatly Regressed Germinal Vesicle, Degenerating Large Follicles and Disturbed Interstitials Cells

Conclusion

From the present study it has been concluded that the doses of testosterone propionate stimulates the spermatogenesis in male rain quail. Similarly the treatment of hydroxyl progesterone caproate in female stimulates the secretion of FSH of pituitary and development of ovarian follicles.

References

Abalain, J. H., Amet, Y. Dahiel, J. Y. and Floch, H. H. (1983). Testosterone stimulation of DNA- Dependant DNA polymerase activities in the cloacal and uropygial glands of the male quail (*Coturnix coturnix japonica*). Gen. Comp. Endocrinology. 52, 164-171.

Alison, M. R. and Wright, N.A. Testosterone 5 alpha reductase activity as related to proliferative status in mouse accessory sex glands: J. endocrinology; 81; 83-92, 1979.

Ambadkar, P. M. and Chauhan, B. C. (1980). Observations on influences of pinealectomy on certain aspects of testicular functions in Indian house crow (*Crovus splendens*). Pavo 18, 39-52.

Arneja, D.V., George, G.C., Mehta, S.N. Dixit, V.P. and Razdan, M. N. (1984). Plasma testosterone concentration in relation of thyroid status in Japanese quail. Indian J. Poultry Science, 19, 207-209.

Arneja, D.V., Sharma, D. W. and Razdan, M.N. (1985). Histomorphology of parathyroids in white perkin drakes. Ind. Journal Poultry Sci. Vol.20, 58-60.

Bently, P. J. (1976). Comparative invertebrate endocrinology, Cambridge University Press, Cambridge, London, New York.

Bhatia, A.J. and Wade, G.N.(1989). Progesterone can either increase or decrease weight gain and adiposity in ovariectomized Syrain Hamsters Physiology and Behaviour, 46:273-278.

C. Berg, L. Holni, I. Brandt and B. Brunstrom, (2001) Anatomical and histological changes in the oviduct of Japanese quail *coturnix* japonica, after embryonic exposure to ethynyloesstradiol. Reproduction 121, 155-165.

Charde, P. N. (1998). Influence of Exogenesis Pharmacological Compounds on Reproductive Biology of *Coturnix coromandelica*. Thesis submitted for Ph.D. Degree Nagpur University, Nagpur.

Follet, B. K. (1984a). In Marshal's Physiology of Reproduction. Vol. 1: Reproductive Cycles of Vertebrates (G.E. Lamming Ed.), pp.283-350.

Fuladi, R. S. (1992). Histopathological and biochemical effects of some drugs on gonads and endocrine glands of grey quail, *coturnix coturnix*. Thesis submitted for Ph.D. Degree, Nagpur University, Nagpur.

Gorbman, A. and Bern, H. A. (1962). A textbook of comparative endocrinology, John Wikey and Sons, New York.

Hervey, E. and Hervey, G. R. (1967). The effects of progesterone on body weight and composition in the rat. Journal of Endocrinology. 37: 361-384.

M.A.Ottinger, S. Pitts and M.A. Abdelnabi (2001). Steroid hormones during embryonic development in Japanese quail; plasma gonadal, and adrenal levels. Paultry Science 8; 795-799.

N. Aste, G.C. Panzica, C. Viglietti. Panzica, N. Harada, J. Balthazar (1998). Distribution and effects of testosterone on aromatase mRNA in the quail forebrain. A non-radioactive *in situ* hybridization study. Journal of Chemical Neuroanatomy 14, 103-115.

2013, Impact of Global Climate Change on Earth Ecosystems *Pages 49–61*
Editors: **D.R. Khanna, A.K. Chopra, Gagan Matta, Vikas Singh & Rakesh Bhutiani**
Published by: **BIOTECH BOOKS, NEW DELHI**

Chapter 7
Conservation of Medicinal Plant *Rauwolfia tetraphylla* (Linn.)

S.G. Desai
Department of Botany, Sevadal Mahila Mahavidyalaya, Nagpur – 440 009, M.S.

Rauwolfia tetraphylla has been introduced in India 150 years ago. This species now grows in wild in seminaturalised condition at various places in India. The plant is perennial and has been identified on the basis of major morphological characters. The fresh seeds do not germinate. It shows innate type of seed dormancy. It takes long period to germinate and eighteen hours photoperiod is suitable for germination. The best germination percentage is examined in one year old seeds. For phytochemical investigations the roots, stem and leaves are collected, washed, dried and finely powdered. The present work shows the presence of various types of phyto-chemicals like alkaloids, flavonoides and steroids in different plant parts. Highest quantity of reserpine is present in leaves. Amino acids are present in all parts of the plant, coumarin is in root; flavonoides, steroids in stem and leaves where as saponins are absent in all parts of the plant. Alkaloids, flavonoides, steroids are present in leaves. Leaves can be plucked and prevent the uprooting of the plant. The same plant can be used year after year which saves cost and energy. The leaves of *R. tetraphylla* are the best substitute for *R. serpentina* which has been over exploited and now become endangered. The immediate need therefore is to educate people in the basic principles of cultivation of medicinal plants so that they may help in managing and to keep environment healthy and make it sustainable. Thus we can conserve biodiversity of nature.

Keywords: Biodiversity, Conservation, Phytochemical, Alkaloid, Rauwolfia.

Introduction

R. tetraphylla (Linn) is a tropical bush and it was introduced into India over a century ago, apparently from Jamaica (Voight, 1845) and has established firmly in the ruderal vegetation. It grows wild in few states of India. It is a perennial plant which flowers and fruits throughout the year provided temperature and moisture conditions are adequate (Rao, 1956) (Figure 7.1).

Rauwolfia has been known for hundreds of years for its medicinal properties and attracted the attention of scientists which belong to various disciplines such as Botany, Chemistry, Biochemistry and Pharmacology. The discovery of tranquilizing action of reserpine stimulates research on *Rauwolfia* and alkaloids present in this plant occupy an important position in drug therapy.

Plant Analysis

Rauwolfia tetraphylla (Linn) is selected for phytochemical screening considering its medicinal properties.

Material and Methods

Phytochemical Screening

The plant parts used were leaves (L), stems (S), roots (R) and fruits (F). The plant material was washed thoroughly and chopped into small pieces and then shade dried. The dried material was grounded to fine powder and stored in polythene bags at room temperature.

TLC of some of the very important phytochemicals such as alkaloids, amino acids, coumarines, flavonoids, saponins, steroids and pigments (anthocyanins and carotenoids) were done at wavelength UV 254 nm and UV 365 nm. These plant samples were screened to detect different classes of secondary metabolites like alkaloids, amino acids, coumarines, flavonoids, saponins, steroids and pigments (anthocyanins and carotenoids) etc; responsible for the medicinal properties. For this experimental procedure is as follows:

Experimental Procedure (Figure 7.2)

Preliminary phytochemical screening of plant was done according to the standard procedure adopted by the pioneer workers (Das and Bhattacharjee, 1970; Gibbs, 1974; Harborne, 1998; Chabra *et al.*, 1993).

Observations

The preliminary screening and TLC of medicinal plant *R. tetraphylla* was undertaken. This screening gave the idea of the chemical compounds responsible for the medicinal properties of the plant. All the phytoconstituents had effect on physiology of the human system, alleviating the human suffering.

Confirmatory Phytochemical Screening

Presence (+) or absence (-) of the various phytochemicals in the different plant parts is given in the Table 7.1 and Figures 7.3A–C and 7.4A–D. The spots were

A Flowering Twig

Plant Bearing Fruits

Figure 7.1: *Rauwolfia tetraphylla*

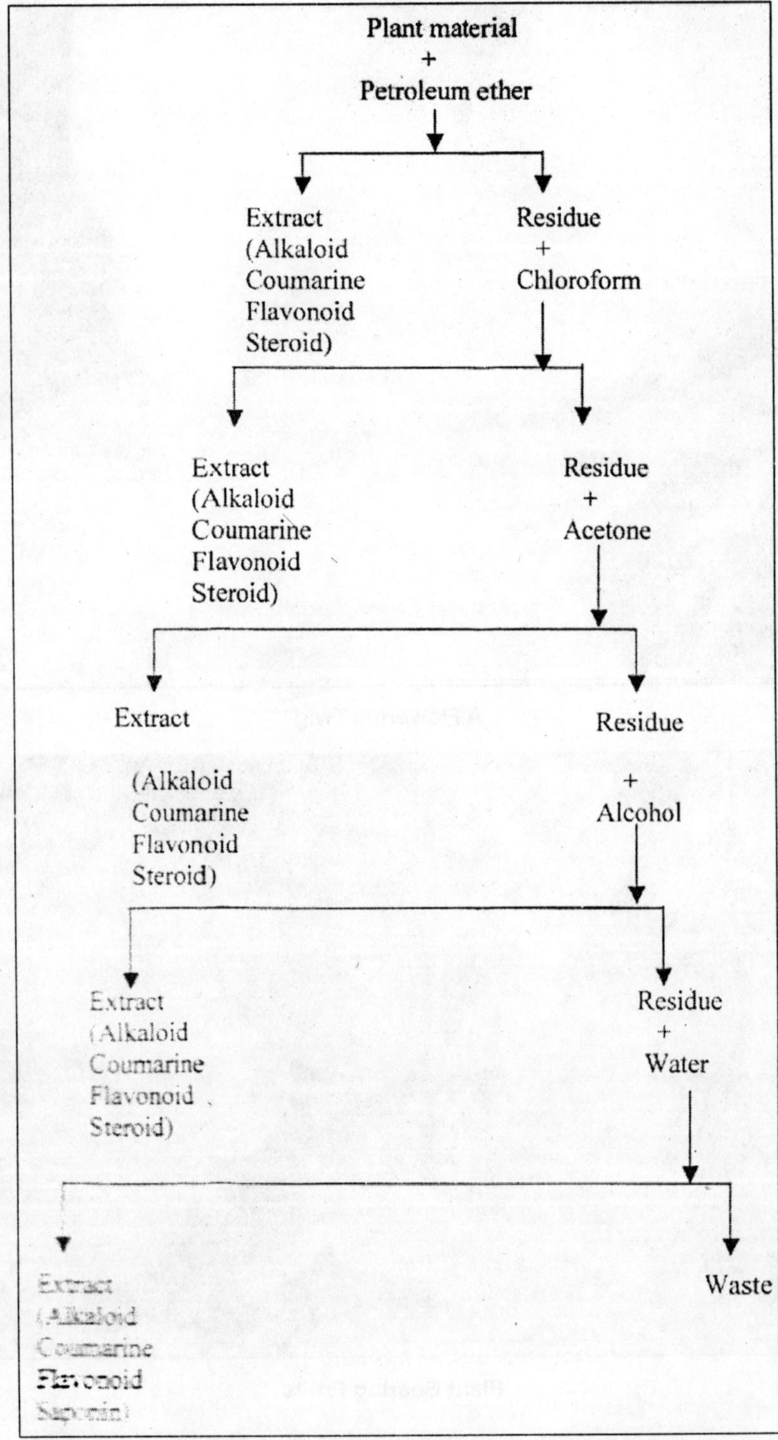

Figure 7.2: Scheme for Preparation of Extract from Plant Material

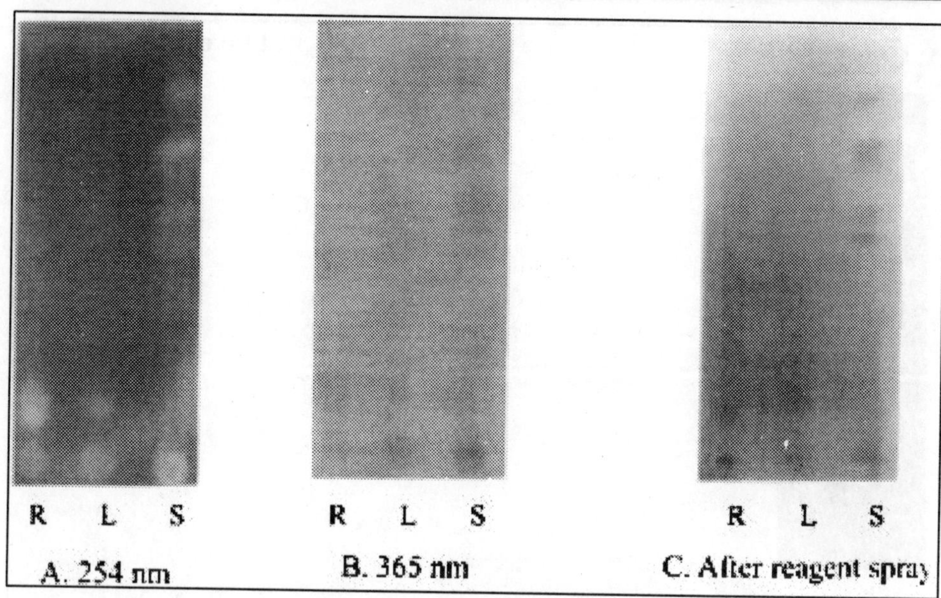

Alkaloids of *Rauwolfia tetraphylla*

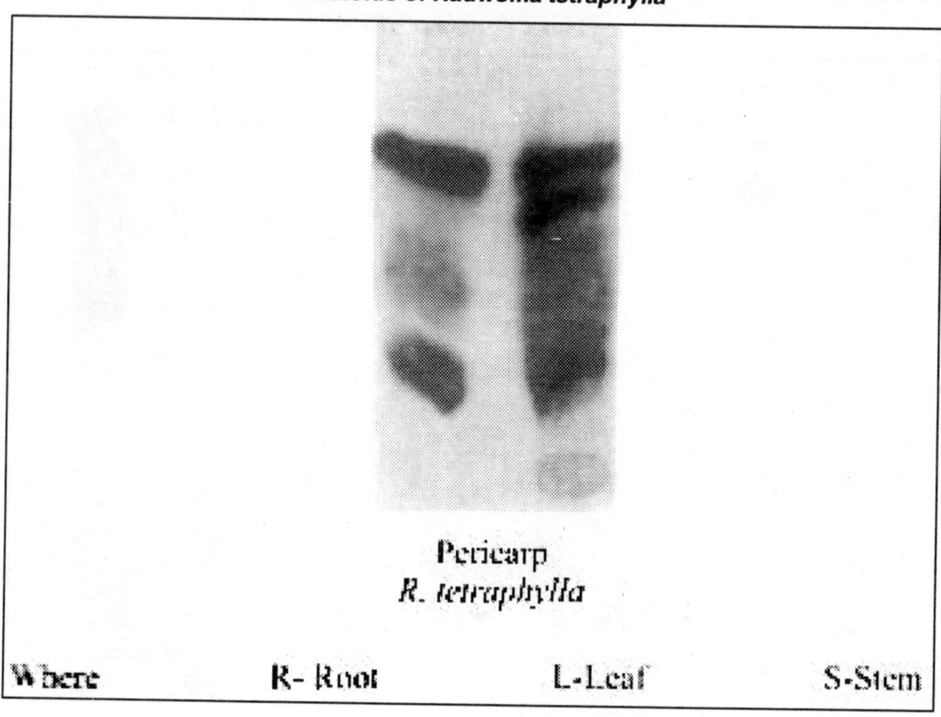

Amino Acids of *Rauwolfia tetraphylla*

Figure 7.3A–C

A. Coumarine
Root
B. Flavonoids (yellow)
R S L F

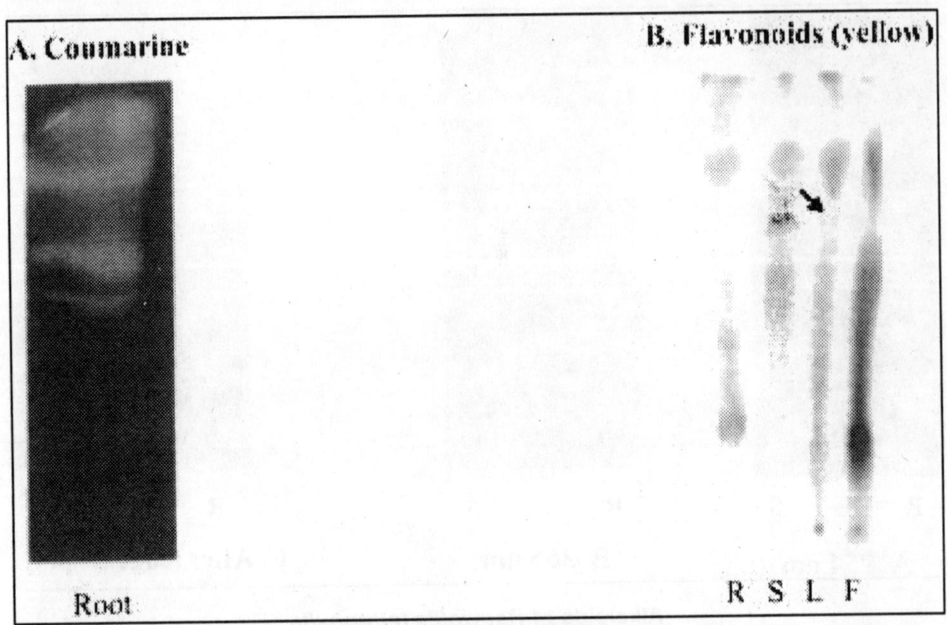

Rauwolfia tetraphylla

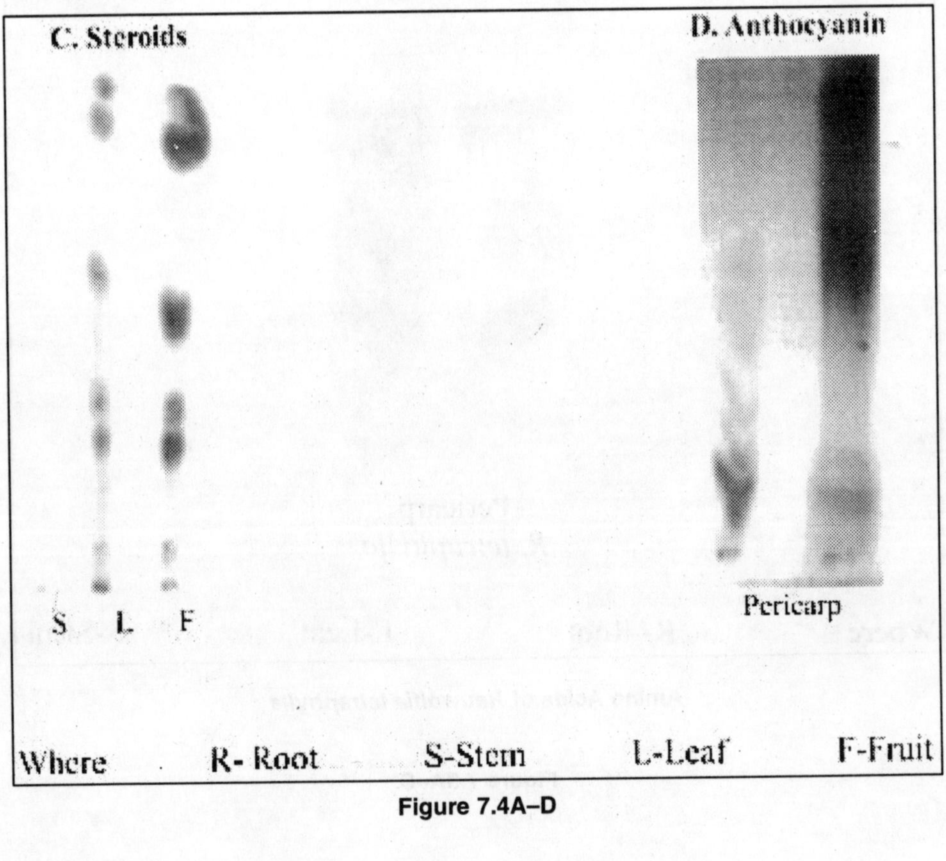

C. Steroids
S L F
D. Anthocyanin
Pericarp
Where R- Root S-Stem L-Leaf F-Fruit

Figure 7.4A–D

compared with Rf values of alkaloids, amino acids, coumarines, flavonoids, saponins, steroids and pigments (anthocyanins and carotenoids). The Rf values of each spot was calculated and noted (Table 7.2).

Table 7.1: Observation Table of Phytochemicals in *Rauwolfia tetraphylla*

Phytochemical	*Reagent*	*Colour*	*Rauwolfia tetraphylla*			
			R	*S*	*L*	*F*
Alkaloid	DRG	Brownish red	+	+	+	–
	Mayer's	Cream ppt	+	+	+	–
	Wagner's	Brown ppt	+	+	+	–
	Hager's	Orange yellow	+	+	+	–
Amino acid	Ninhydrin	Violet purple	+	+	+	+
Coumarine		Bluish-Green fluorescence	+	–	–	–
Flavonoid	Shirnoda	Pink or magenta red	–	+	+	–
Saponin	Foam formation	Foam	–	–	–	–
	Libermann Buchard	Violet	–	–	–	–
	Hesse's	Pink	–	–	–	–
Steroid	Salkawski	Greenish yellow	–	+	+	+
	Libermann Buchard	Red	–	+	+	+
Pigment						
Anthocyanin		Red	–	–	–	+
Carotenoid		Blue-green	–	–	–	–

Table 7.2: Rf Values of Detected Phytochemicals in *Rauwolfia tetraphylla*

Phytochemicals	*Parts of the Plant*	*Rf Values*
Alkaloids	Root	0.12, 0.46, 0.53
	Stem	0.14
	Leaf	0.52, 0.55, 0.71, 0.85
	Fruit	–
Amino acids	Root	0.4, 0.5, 0.6, 0.75
	Stem	0.2, 0.48, 0.57
	Leaf	0.34, 0.4, 0.5, 0.65
	Fruit	0.2, 0.3, 0.55, 0.7
Coumarines	Root	0.55, 0.65, 0.63, 0.91
	Stem	–
	Leaf	–
	Fruit	–

Contd...

Table 7.2–Contd...

Phytochemicals	*Parts of the Plant*	*Rf Values*
Flavonoids	Root	–
	Stem	0.7
	Leaf	0.5, 0.6, 0.68
	Fruit	–
Saponins	Root	–
	Stem	–
	Leaf	–
	Fruit	–
Steroids	Root	–
	Stem	0.29, 0.25, 0.67
	Leaf	0.08, 0.29, 0.91, 0.98
	Fruit	0.27, 0.52, 0.85, 0.96
Pigments		
Anthocyanin	Fruit	0.22, 0.56
Carotenoid	Fruit	–

Discussion

Phytochemical screening is of paramount importance in identifying new sources of therapeutically and industrially important compounds having medicinal importance like alkaloids, amino acids, coumarines, flavonoids, saponins, steroids and pigments etc. To make the best and judicious use of available natural wealth, a number of medicinal plants have been chemically investigated. The active principles isolated from them are used as medicines. Kato *et al.* (2002), Biswas and Rahman (2008).The discussion is based on Table 7.3. The phytochemical analysis of *R. tetraphylla* is discussed under three categories (Harborne, 1998).

1. Terpenoids
2. Phenolics
3. Nitrogenous compounds

Triterpenoids is a major phytochemical group of active principles in medicinal plants. Triterpenoids can be divided into at least four groups of compounds: true triterpenes, steroids, saponins and cardiac glycosides. The later two groups are essentially triterpenes or steroids which occur mainly as glycosides.

Steroids are also included as derived triterpenes. Steroids are well known compounds for their activity as sex hormones.

One of the glycoside is saponins which are soluble in waste and form an emulsion due to their hydrophobic character. Cardio glycosides are found mostly in Apocynaceae (Daniel, 2005). Saponin of marked pharmacological activity is obtained

Table 7.3: Phytochemistry of Different Species of *Rauwolfia*

Name of the Plant	*Part of the Plant*	*Author*	*Chemical Constituent*
R. vomitoria		Ronchetti *et al.* (1971)	Sandwicine and Isosandwicine
R.sandwicenis		Ronchetti *et al.* (1971)	Sandwicine and Isosandwicine
R.callra	Root	Habib and Court (1973)	Indole alkaloids *viz.*, Acricine, Renoxidine, Sarpegine and Yohimbine
R.macrophylla	Root	Timmens and Court (1974)	Indole alkaloid
R.obscura	Root	Timmens and Court (1976)	Diester-alkaloid-Reserpine, Rescinnamine
R.obscura	Stem	Timmens and Court (1976)	Alkaloids
R.reflexa		Chatterjee *et al.* (1978)	Indole-alkaloids-Rauflexine, Reflexine
R.cumminsii	Stem	Iwu and Court (1978)	Indole-alkaloids
R. psychotrioides		Hernan *et al.* (1979)	Rauvoxinine and Triterpenes
R. oreogiton	Leaves	Akinloye and Court (1980)	Indole-alkaloids
R. vomitoria	Leaves	Amer and Court (1980)	Indole-alkaloids
R. volkensii	Leaves	Amer and Court (1981)	Indole-alkaloids
R. serpentina	Root	Siddiqui *et al.* (1987 a)	Ajmalacine
R. serpentina	Root	Siddiqui *et al.* (1987 b)	Yohambinine
R. tetraphylla and *R. cubana*	Root	Martinez *et al.* (1989)	Sperpagine
R. serpentina	Culture	Endref *et al.* (1993)	Raumaelines and other alkaloids
R. serpentina	Culture	Benjamine *et al.* (1994)	Ajmaline, Serpentine
R. grandiflora	Bark	Bianco *et al.* (1994)	Monoterpenoid (Iridoids)
R. sprucci		Madinaveitia *et al.* (1995)	Indole-alkaloid
R. sellowii		Batişta, *et al.* (1996)	Sellowine
R. serpentina	Root	Warzecho *et al.* (1999)	Reserpine
R. nitida	Root	Amer and Court (1981)	Raucaffricine-o-β, D-Glucoside. Reserpine, Serpetinine, Pseudoreserpine, Reserpiline.
R. vomitoria	Stem	Sabri and Court (2001)	Alkaloids
R. radix		Jin *et al.* (2002)	Reserpine
R.bahiensis		Kato *et al.* (2002)	13 indole alkaloids.
R. tetraphylla	Root, Stem, Leaves, Fruit	Present study (2009)	Alkaloid, Amino acid, Flavonoids, Steroids, Anthocynin and Carotenoids

from a number of plants. Chinese system of medicine is particularly rich in saponin drugs. Saponins are well known expectorant. In the present investigation it was observed that saponins are completely absent in all the parts of *R. tetraphylla* (Table 7.2).

Coumarines are aromatic lactones. They show a number of clinical properties ex. diuretic, antispasmodic, oestrogenic, hypnotic, anticholerostatic, antiarterosclerotic and vasodilatory. Flouro-coumarine is used as skin tanner. The animal coumarine is used as a flavouring agent in cigarettes.

Coumarines are present only in the roots of *R. tetraphylla*, which are anticoagulant, estrogenic, vasodilators, antibacterial and antihelminthic in property. So the plants reported positive for this compound can be used in drugs as anticoagulant and vasodilators (Table 7.2).

Flavonoids are mainly water soluble compound. Brun *et al.* (1999) established a new flavonol glycoside from stem of *Catharanthus roseus*. This they named as syringelin. According to them this is the first report about isolation of this phytochemical in Apocynaceae. Another flavonol, sinapoyl glycoside is present in the leaves of *Thevetia peruviana* (Abe *et al.*, 1995) plants with flavonoid content can also be used on rheumatic fever and arthritis (Tripathi and Rastogi, 1981).

Flavonoids are present in stem and leaf of *R. tetraphylla* (Table 7.1). Species of *Rauwolfia* are rich source of bio-active indole alkaloid, such as Reserpine in the WHO Model list of Essential Drugs as an antihypertensive agent.

Phytochemical screening shows that alkaloids are present in roots, stem and leaves of *R. tetraphylla* (Table 7.2). Highest quantity of this compound is reported in the leaves of *R. tetraphylla*. D.A.A. KIDD and P.G.W Scott (12th April 2011) they studied the distribution of reserpine in several two phase system and also reported the presence of alkaloids in several species of *Rauwolfia*.

Rauwolfia is the known for the roots from which a drug reserpine is prepared which lowers the blood pressure. From the comparative Table 7.3, it is seen only 4 species of *Rauwolfia*, whose leaves show the presence of the alkaloids including *R. tetraphylla*. R Harisaranraj, K. Suresh and Saravanababu (2009) also reported the presence of bio active alkaloids, saponins, flavonoids and minerals in *Rauwolfia serpentina* and *Ephedra vulgaris*. These are R. *oreogiton*, *R. vomitoria* and *R. volkensii*. *R. tetraphylla* is less toxic. Mokutima A. Eluwa and *et al.* (2010). In their comparative study of teratogenic potentials of crude ethanolic root bark and leaf extract of *R. vomitoria* on the fetal heart, they observed the high doses of ethanolic leaf and root extracts may be cardiotoxic to the developing rat's heart. It is used as an anti-poisoning agent for snake, spider and scorpion and also anti-dot for fever, worms and pimples.

The present phytochemical studies on *R. tetraphylla* also show the presence of alkaloids in the leaves. The harvesting of the leaves is a better source that the roots. Individually, the leaves can be plucked, without disturbing the main plants while for the roots, the plant is to be totally uprooted. In *R. tetraphylla* the same plant can be used year after a year, which will be cost and energy saving. In the present day circumstances, the herbalist must come forward by growing of more plants of *R. tetraphylla* as compared to *R. serpentina* as a non destructive exploitation. There are other species of *Rauwolfia* also which produce alkaloids, which is given in the Table 7.3.

Conclusion

The present phytochemical studies showed that alkaloids are present in leaves. The harvesting of leaves is a better source than roots. Leaves can be plucked and prevent the uprooting of the plant. *R. tetraphylla* the same plant can be used year after year which saves cost and energy. The use of *R. tetraphylla* become the best substitute for *R. serpentina* which has been over exploited and now become endangered.

Futuristic Approach

- ✰ Mass scale cultivation of *Rauwolfia tetraphylla* be encouraged for alkaloids from leaves.
- ✰ The plant can be grown in open coal mines, minor irrigation bands and round about villages.
- ✰ Coumarine can be used in allopathic drugs.

Acknowledgement

I am thankful to Dr. Pravin Charde, Principal, Sevadal Mahila Mahavidyalaya, Nagpur, for providing the research facilities. I am also thankful to my colleagues, Dr Dongre, Dr. (Mrs.) Kulkarni and Mr. Lambat for their cordial co-operation.

References

ABE, F., YAMAUCHI, T., YAHARA, S. AND NOHARA, T., 1995. Minor iridoids from *Thevetia peruviana* Phytochemistry, 38 (3) 793-794.

AKINOLOYE, B.A. AND COURT, W.E., 1980. Leaf alkaloids of *Rauwolfia volkensii* Phytochemistry, 19 (2) 307-311.

AMER, M.A. AND COURT, W.E., 1980. Alkaloids of *Rauwolfia vomitoria.* Phytochemistry, 19 (8) 1833-1836.

AMER, M.A. AND COURT, W.E., 1981. Alkaloids of *Rauwolfia nitida* root bark. Phytochemistry, 20 (11) 2569-2573.

BATISTA, C.V.F., SCHRIPSENA, R.B. AND HENRIQUES, A.T., 1996. Indole alkaloids from *Rauwolfia sellowii* Phytochemistry, 41 (3) 969-973.

BENJAMIN, B.D., RAJA, G. AND HEBLE, M.R. 1994. Alkaloids synthesis by root cultures of *Rauwolfia serpentina* transformed by *Agrobacterium rhizogenes* Phytochemistry, 35 (2) 38 1-337.

BIANCO, A., LUCA, A.D., MAZZCI, R.A. AND LIMA, R.A. 1994 Iridoids of *Rauwolfia grandiflora* Phytochemistry, 35(6) 1485-1487.

BISWAS, S., AND REHMAN, I. 2008, Curcumin, an active ingredient of Haldi, American Journal of Respiratory Cell and Molecular Biology, Vol.39, pp.312-323.

BRUN, G., DIJOUX, M.G., DAVID, B., MARIOTTE, A.M., 1999. A new flavonol glycoside from *Catharanthus roseus* Phytochemistry, 50 167-169.

CHATTERJEE, A., CHAKRABARTY, M. AND GHOSH, A.K., 1978. Indole alkaloids of *Rauwolfia reflexa*. The structures of rauflexine and reflexine. Tetrahedron Letters, 19(40) 3879-3882.

CHHABRA, S.C., VISO F.C. AND MSHIU, F.N., 1993. Phytochemical screening of Tanzanian medicinal plants J Ethnopharmacology, 11 157-179.

DANIAL, M., 2005. Herbal Technology: Concept and Scope, Narosa publications, New Delhi.

D. A. A KIDD and P. G. W. SCOTT Alkaloids of *Rauwolfia* species. Journal of pharmacy and pharmacology article published online 12 April 2011.

DAS, A.K., AND BHATTACHARJEE, A.K., 1970. A systemic approach to phytochemical screening Tropical Science XII 54-58.

ENDREF, S., TAKAYAMA, H. AND STOCKIGT, J. 1993. Alkaloids from *Rauwolfia serpentina* cell cultures treated with ajmaline Phytochemistry, 32(3) 725-730.

GIBBS, R.D., 1974. Chemotaxonomy of Flowering Plants I-TV Montreal and London McGill Queens University Press.

HABIB, M.S. AND COURT, W.E., 1973. Minor alkaloids of *Rauwolfia caffra*. Phytochemistry, 12 (7) 1821.

HARBORNE, J.B., 1998. Phytochemical Methods, A Guide to Modern Techniques of Plant Analysis. Chapman and Hall, London (Second Edition).

HERNAN, E., CARDOVA, B. AND PEFIA, C.A., 1979. Alkaloids from *Rauwolfia psychotrioides* Phytochemistry, 18 (8) 1419–1420.

IWU, M.M. AND COURT, W.E., 1978. Alkaloids of *Rauwolfia cumminsii* stem Phytochemistry 17(9) 1651-1654.

JIN, G.B., HONG, T., INOUE, S., URANO, T., CHO, S., KOJI, 0., MAYA, K., OUCHI, Y. AND CYONG, J.C., 2002. Augmentation of immune cell activity against tumor cells by *Rauwolfia radix* J Ethnopharm 81(3) 365-372.

KATO, L., BRAGA. R.M., KOCH, I. AND KINOSHITA, L.S., 2002. Indole alkaloids from *Rauwolfia bahiensis* A.DC. (Apocynaceae). Phytochemistry, 60 315-320.

MADINAVEITIA, A., VALENCIA, E., BERMEJO, J. AND GONZALEZ, A.G. 1995. Indole alkaloids from *Rauwolfia sprucci*. Biochemical systematics and Ecology, 234 (7-8) 877.

MARTINEZ, J.A., VELEZ, H. AND SANTANA, T., 1989. An alkaloids from two *Rauwolfia* sp. Phytochemistry 28 (3) 961-961.

MOKUTIMA A ELUWA and *et al.*, Ph. D thesis 2010. Comparative study of teratogenic potentials of crude ethanolic root bark and leaf extract of *Rauwolfia vomitoria* on the fetal heart.

RAO, A.S. 1956. A revision of *Rauwolfia* with particular reference to the American species. Ann Mol Bot Gard, 43 253–353.

RONCHETTI, F., RUSSO, G., BOMBARDELLI, E. AND BONATI, A. 1971. A new alkaloids from *Rauwolfia vomitoria* Phytochemistry, 10 (6) 1385-1388.

R. HARISARANRAJ, K SURESH and SARAVANABABU. Advances in biological research 3 (5-6): 174-178, 2009. Evaluation of the chemical composition *R. serpentina* and *Ephedra vulgaris.*

SABRI, N.N. AND COURT, W.E. 2001. Stem alkaloids of *Rauwolfia vomitoria* Phytochemistry, 17 (11) 2023-2026.

SIDDIQUI, S., AHMED, S., HAIDER, S.I. AND SIDDIQUI, B.S. 1987a Ajmalicine an alkaloid from *Rauwolfia serpentina* Phytochemistry, 26 (3) 875-877.

SIDDIQUI, S., AHMED, S., HAIDER, S.I. 1987b Isolation of a new alkaloid "yohambinine" from *Rauwolfia serpentina* Benth. Tetrahedron Letters, 28 (12) 1311-1312.

TIMMINS, P. AND COURT, W.E. 1976. Root bark alkaloids of *Rauwolfia obscura* Phytochemistry, 13(9) 1997-1998.

TIMMINS, P. AND COURT, W.E. 1974. Alkaloids of *Rauwolfia macrophylla* Phytochemistry, 13(2) 281-282.

TIMMINS, P. AND COURT, W.E. 1976. Stem alkaloids of *Rauwolfia obscura* Phytochemistry, 15(5) 733-735.

TRIPATHI, V.D. AND RASTOGI, R.P. 1981. Flavonoids I biology and medicine J Sci lnd Res, 40 116-124.

TREASE, G.E. AND EVANS, W.C. 1954. *Rauwolfia serpentine* and other species Pharm Jour., 172 351-353.

VOIGHT, J.O., 1845. Hortus Suburbanus Calcuttensis, p.532.

WARZECHA, H., OBITZ, P. AND STOCKJGT, J. 1999. Purification, partial amino acid sequence and structure of the product of raucaffricine-O-β-D- glucosidase from plant cell cultures of *Rauwolfia serpentina* Phytochemistry,50 1099–1109.

2013, Impact of Global Climate Change on Earth Ecosystems *Pages 63–74*
Editors: **D.R. Khanna, A.K. Chopra, Gagan Matta, Vikas Singh & Rakesh Bhutiani**
Published by: **BIOTECH BOOKS, NEW DELHI**

Chapter 8
Diversity and Nesting Habit of Wasps (Akola City)

Mangala B. Karlekar
Shri Shivaji College of Arts, Commerce and Science, Akola, M.S.

There are more than 1, 00,000 species of wasps crawling and flying all over the earth and large of them are still undiscovered due to severe parasitic mode of life. So, it comes as no surprise that a great variety exists when it comes to their colour, size, shape and life style. They have very diversified behaviour, habit, nest building, caste system, chemical communication systems, etc. During the course of study 20 species of paper wasps and mud wasps were found at the 7 study sites. Wasps are ecologically important as flowers and wasps share a mutually beneficial relationship. Wasps may not produce honey, but wasps perform a vital service by helping to pollinate the world's plant life and eliminate various six- and eight-legged pests that feed on crop. Wasps were collected from various areas in and around Akola city and identified by using identification key characters. There are two types of wasps, social wasps and solitary wasps. Social wasps are insects that tend to thrive in colonies while solitary wasps live alone usually building a single nest made out of clay or mud and demonstrate defensive behaviours by penetrating sclerotized sting into the victim. Wasps are Biocontrol agents many species of wasps inhabiting in agro eco systems play a crucial role in controlling pest populations. All species of wasps are predatory, and they may consume large numbers of flies, caterpillars, bugs, spiders by preying on them. This really does help to cut down of use of harmful pesticides in the long run. In some places in the world people use the immature stages (Larvae& pupae) of wasps as food for good source of protein. The main motto of wasps diversity study to assist in increasing the awareness about diversity and importance of the wasps as part of worlds bio wealth.

Keywords: *Social insects, Biological control, Venom, Wasp.*

Introduction

About 100 million years ago, during the cretaceous period, the planet earth was devoid of flowering plants and occupied mostly by conifers. This evergreen world crawled with insects such as Ants and their winged cousins the wasps (both members of order *Hymenoptera*) (Lamb, 1998). In the world of entomology, wasps (are some of the most feared of all insects because of their stings. Wasps live on all continents inhabited by people. There are more than 1,00,000 species of wasps crawling and flying all over the earth and large of them are still undiscovered due to severe parasitic mode of life. So, it comes as no surprise that a great variety exists when it comes to their colour, size, shape and life style. They may be free-living, phytophagous, predatory, entomophagous, parasitic and social insects. They have very diversified behaviour, habit, nest building, caste system, chemical communication system, etc. (Tembhare, 2006). Wasps are known for their stings. Stingers are effective weapons because they deliver venom that causes pain when injected into the skin. The stinger is a modified egg-laying apparatus. So, only females can sting. Wasp's venom is produced inside a venom gland, from here, it seeps out through a barbless stinger.

Flowers and wasps share a mutually beneficial relationship. Wasps may not produce honey, but wasps perform a vital service by helping to pollinate the world's plant life and eliminate various six-and eight-legged pests that feed on crop (Lamb, 2008).

There are two types of wasps, social wasps and solitary wasps. Social wasps are insects that tend to thrive in colonies while solitary wasps live alone usually building a single nest made out of clay or mud and demonstrate defensive behaviors that sometimes are harmful to humans. In terms of success, however, social wasps have earned due respect as major influences and elements of most ecosystems. Paper wasps build clusters of hexagonal paper cells. Mixing masticated (chewed) wood pulp with adhesive saliva, these paper nest cells act as larval nesting chambers for the young wasps. The paper nest may be bag-like (calyptodomous) and contains a number of tiers of cells (*e.g.* in hornets and yellow jackets colonies) or open (gymnodomous), with a single tier (as in *polistine* species) with hexagonal shaped cells (Hermann, 2009). The subfamily Polistinae is a diverse group of eusocial wasps, being represented in the New World by 21 genera (Carpenter, 2004).

Ecology: Wasps are Highly Important to Ecosystems

Wasps feed on flower nectar and play a role in pollination. Although most people think of wasps as pests they benefit mankind and environment in many ways. They help control arthropods (flies, caterpillars, bugs, spiders, etc.) by preying on them.This is healthier for our environment in the long run. Wasps are also used in research and experiments. Many parasitic wasps have been cultured and used in the biological control of agricultural pests. A considerable amount of research has been carried out to determine if *polistine* species can be used in field settings to control pestiferous caterpillars.

Purpose of Study

Almost every pest insect species has at least one wasp species that preys upon it or parasitizes it, making wasps critically important in natural control of their numbers, or natural biocontrol. Most other wasps are either parasitic or predaceous and therefore play a vital role in limiting the populations of thousands of other insect species. Social wasps are sometimes said to be beneficial insects, ridding gardens of insect pests. Potter wasps are important in the natural control of caterpillars. All wasps are eaten by other species, thereby providing many links in the food web.

The knowledge of Indian wasps is scanty and fragmentary. Such valuable and significant benefits of wasps to human far outweight the harm which they do, prompted me to study about wasps.

Materials and Methods

Geographical and Climatic Conditions of Akola District

The location of Akola city is latitude: 20.7° North and longitude: 77.07° East. Annual temperatures range from a high of 48 °C (118 °F) to a low of 10 °C (50 °F). Akola is very hot during summer season, especially in May. But even Akola city is located at an altitude of 925 ft. (282 m.) above the sea-level. The climate is tropical. Annual temperatures range from a high of 48 °C (118 °F) to a low of 10 °C (50 °F).

The annual rainfall averages 34 inches. Most of the rainfall occurs in the Monsoon season between June to September, but some rains can also be seen during January and February. On the north, Akola district is bounded by the Melghat Hills and Forest region. Here are some of the rivers in Akola, with their tributaries: Purna River, Uma River, Katepurna River, Shahanur River, Morna River, Mann River, Aas River, Vaan River, etc.

Collection of Wasps

The wasps were collected by sweep net from various places within and around Akola region such as Shastri Nagar, Ganga Nagar, Farm at Kalmeshwar Road, A1 Sawmill, Dr. Panjabrao Deshmukh Krishi Vidyapeeth, Akola Campus; Shri Shivaji college of arts commerce and Science College Campus, Akola and Patur. The wasps were kept in small box in the refrigerator for 30-50 minutes, so that they become inactive and can be easily handled for observation. Then wasps were killed by ethyl acetate and preserved as dry specimen and in 70 per cent alcohol.

Wasp's Identification Guides

Identifications of wasps were done as per the dichotomous keys or diagnostic characters available on the internet websites, along with the pictures of each species on google seach images, *e.g.* Wikipedia, the free encyclopedia, Natural History Photographs by Cor Zonneveld. "Indian Institute of Science

Observations and Results

The paper wasps are classified under Kingdom - Animalia. They belong to the Phylum - Arthropoda (having exoskeletons and jointed appendages) which includes Class - Insecta (insects having a head, thorax and abdomen and six legs). The class is further sub divided into Order - Hymenoptera (this order includes all ants, bees, and wasps). Under this categorized is the Family – Vespidae. The vespids are a family of wasps, including all social wasps (hornets, paper wasps, yellow jackets.etc) and some solitary wasps. This also consists Subfamily: Polistinae (generally referred to as polistines). In our observations we find five species of subfamily polistines and thethree species of Subfamily: vespinae (generally referred to as vespines) (Hermann, 2009).

In our investigation following species of family vespidae, subfamily-*Polistene sp.* were found *Ropalidia marginata, R. variegata, R. flavopicta, Polistes stigma, Polistes humilis.* Family-Vespidae, Subfamily-Vespinae Genus and Species found were *Vespa tropica and Vespa orientalis,* Family-Vespidae, Subfamily-Eumenidae, spp. found were *Delta esuries and Delta conoideum.*Family Vespoidea- *Rhynchium atrum, Rhyncium brunneum.* Family-Sphecidae (Thread waisted wasp)Subfamily-Sceliphrinae, species found were *chalybion californicum and Sceliphron caementarium,* few species of Cuckoo wasp belongs to family Chrysididae, subfamily chrysidinae, species found were *Stilbum cynaurum,* and Subfamily parnopidae.*Parnopes* species was found. Family- pampilidae (Spider hunting wasp) Subfamily-pepsinae, species found were *Hemisepsis* (Tarantula Hawk wasp), Family-Ichnemonidae family- Braconidae. The no. is large, hence figures and identified characters of some species was given in tabulated form.

Discussion

A wasp has the body of a predator, not a forager. Social wasps are hunters and scavengers that prey on a variety of arthropod and animal protein sources. The effectiveness of wasp stings and venoms as defences has allowed wasps to evolve bright warning colours of red, yellow, orange, or white on a black or dark background. These conspicuous warning colour patterns are called aposematic, and their brightness usually correlates with the degree of painfulness of the sting. Potential predators and humans alike learn to avoid beautiful aposematic wasps (Schmidt, 2006).

Akola region being located in the tropics, its climate and seasons do have contribution in establishing a vast diversity of each insect species and so that of wasp. The social wasps or paper wasps are common, flying in a sunny day or in the house garden during early rainy season. The Ropalidia is the more distributed one species here.

The polistes species are confined to only forest and agricultural areas or densely tree packed areas, where they can find much milder weather and food in abundance. The potter or mason wasp appears in the scene, during the early rainy season in Akola region. Till august one can find mud nests build in attics, roof, corners of walls etc.

Table 8.1: Different Species of Wasps Found in and around Akola City

Sl.No.	Species Classification	External Features	Food Source	Nest Location	Ecological Importance
1.	Family: Vespoidea, Tribe: Ropalidiini, Genus: *Ropalidia,* Specific name: *flavopicta* - (Smith, 1857), Scientific name: *Ropalidia flavopicta* (Smith, 1857)	This species is 13mm in body length. Body black colour with reddish brown and yellow striae and bands, Antenna yellowish brown slightly swollen apically, scape slightly curved and weakly flattened dorsoventrally, clypeus nearly flat, bluntly pointed apically, Legs dark ferruginous tibiae and tarsi yellowish brown, wings hyaline, darker along costa; marginal cell not infuscate black smoky patches/ stigma at the distal end of costa (Kojima, 1997) (Figure 1).	Adults forage for nectar, their source of energy, and the larvae are fed on chewed up cater-pillars caught by the adults.	It is well known as a nest defender and may represent the greatest threat to intruders, the nest was built high above the ground on the venti-lation window, paper honeycomb nests. it is stellocyttarous The nest may be one or more years old as it appears to be used multiple times for colony cycle (Figure 4).	They are natural enemies of many garden insect pests, so are considered to be beneficial insects
2.	Family: Vespidae, Sub-family: Polistinae, Tribe Polistinii, Genus: *Polistes,* Species: *humilis*, Scientific name: *Polistes humilis* (*Fabricius*, 1781) Common Name: Australian Paper Wasp	Body slender with narrow waist and mostly brownish-red with thin yellow cross bands on the the spindle shaped abdomen which tapers at both ends and is never truncate at the base, presence of a distinct lobe on the hind-wings and by, thorax is deep chocolate brown with the end of the abdomen reddish-brown, mandibles and clypeus yellow except reddish brown area basally, gaster with 2 or 3 yellow bands. Paper wasp Polistes humilis has a small head, with medium sized eyes and medium length antennae. Wings reddish or amber brown having dark spot near the tip off each forewing, in Polistes the 2nd gastral segment clearly overlaps the 3rd segment (Figure 2)	The wasps prey on lepidopterous larvae to feed their larvae. Adult paperwasps mainly feed on nectar from flowers while the larvae are fed caterpillars and other insect prey.	The nests of most species are suspended from a single, central stalk, or pedicle has the shape of an upside-down umbrella. The Combs are semi-concentric. The nest color is milky white. It commonly builds its nest in trees or bushes, usually on the branches, but sometimes on stems and le.aves. Often nests are hidden inside dense shrubs (Figure 5).	The Polistes wasps are very beneficial predators of caterpillars.

Contd...

Table 8.1–Contd...

Sl.No.	Species Classification	External Features	Food Source	Nest Location	Ecological Importance
3.	Family: Vespidae, Genus *Vespa*, Specific name: *orientalis*, Scientific name: *Vespa orientalis* (Linnaeus, 1771), Common Name: Oriental Hornet	The female queen measures 25 to 35 mm long, males and workers are smaller. In males; Size of the fertile female is upto 28 mm and of sterile female is up to 22 mm. This species is characterized by having brown body, Thorax largely reddish, pronotum, a triangular yellow spot under each wing base, with clypeus, frons, and scape ventrally yellow, Mesoscutum, Scutellum and Post-scutellum reddish brown, Reddish brown or ambar colored hairs on head and thorax short, sparse, stiff and bristle-like (Figure 3).	The Oriental Hornet is a carnivore. It attacks, kills and eats other insects. It also scavenges any meat that it finds. Adult food consists of nectar or other sugary solutions such as honeydew and the juices of ripe fruits.	*Vespa orientalis* builds underground nest. They also chew plant material, like tree bark; to build their nests. The primary nest is built by the queen alone and is known as an "embryonic nest."	These hornets hunt other insects, caterpillars, spiders, and even other bees and wasps to supply protein to larvae.
		The Potter wasp or the mason wasp			
4.	Superfamily: Vespoidea, Family: Eumenidae, Genus: *Delta*, Specific name: *conoideum*, Scientific name: *Delta conoideum* (Gmelin, 1790)	Female: 23-26 mm; male: 18-22 mm. Forewing length: female: 19mm; male: 16mm.: mandibles reddish and, clypeus yellow, pyriforme, antennae are reddish, in female, clypeus and lower half of frons yellow, *conoideum* a broad transverse band across the apex between the tops of the eyes, gaster is dark red, smooth and shining with the surface minutely aciculate, with base of the second tergite and a short transverse medially interrupted band on its middle above black, its apex truncate, legs pale reddish.	The diet of potter wasps consists primarily of nectar, spiders, beetle larvae, and young caterpillars.	The nests are round with a thin neck at the top, much like a cork bottle or special vase. They fly the paralyzed caterpillars to the hive they created. They lay their eggs on these larvae inside nests, one egg in each cell. And then seals off the cells.	

Contd...

Table 8.1–Contd...

Sl.No.	Species Classification	External Features	Food Source	Nest Location	Ecological Importance
		Apical half of second gastral tergite and whole of the remaining tergites Head yellow except thorax dark red with black patches on mesoscutum, forewing longitudinally plaited at rest basally flavohyaline and apically fusco-hyaline, with a slight purplish refringes (Figure 7).			
5.	Family: Vespoidea, Genus: *Rhynchium*, Specific name: *atrum*, Scientific name: *Rhynchium atrum* (Saussure, 1852)	Body entirely black, wings dark bluish black, body size larger than former Rhyncium species. The middle tibia has one apical spur. The claws are forked (bifid). The mandibles are elongate and crossing. Another thing to note about potter wasps is that when at rest, their wings are usually held at an angle, similar to the *Vespid* wasps, instead of folded over their backs at a very narrow or overlapping angle like other solitary wasps such as the *Sphecids* or *Pompillids*, the pronotum and the transverse carina of scutellum have reddish colored tinges, the abdomen is smooth and does not shows punc-tuation as in former Rhyncium species (Figure 6).	Adults feed on flower nectar and collect small caterpillars to feed their brood. The caterpillars are paralyzed with the wasp's sting piled into the brood. The female wasp then lays an egg on the stored caterpillars. The potter wasp larva consumes from 1 to 12 caterpillars as it grow	It builds mud nests with multiple barrel shaped cells each separated from other. The nest consists of 4 to 15 cells (Figure 6).	They are all predators
			Digger wasp		
6.	Family: Sphecidae, Superfamily: Apoidea (traditional Sphecidae),	This beautiful wasp is 12 to 18 mm long, metallic blue, with light copper or golden colored wings, the wings	Adults of this are larvae, feeding on spiders provided by their mother.	*Chalybion* also uses plant stems and cavities for nests. It has been	They are also the main predator of the black and brown widow spiders.

Contd...

Table 8.1–Contd...

Sl.No.	Species Classification	External Features	Food Source	Nest Location	Ecological Importance
	Family: Sphecidae, (Thread-waisted Wasps), Subfamily: Sceliphrinae, Tribe: Sceliphrini Genus: *Chalybion,* Specific name: *californicum,* Common Name: Blue Mud Wasp, Scientific name: *Chalybion californicum* (Saussure, 1867)	of the mud-dauber, as of all the diggers, are not folded fan-like down the middle. Legs are long and slender, and ornamented with spines. Petiole present between posterior region of the thorax and the gaster of the wasp (Figure 8)	Adult females capture orb-weavers and comb-footed spiders often including black widow spiders) or crab spiders. These wasps capture their prey by paralyzing them with a sting (Bohart and Menke 1976, Hogue 1974, O'Brien 1998)	reported that this species does not build its own nests, but occupies those built by *Sceliphron caementarium* (Figure 11).	
7.	Apoidea-traditional Sphecidae), Family: Sphecidae, (Thread-waisted Wasps), Subfamily: Sceliphrinae, Tribe: Sceliphrini, Genus: *Sceliphron,* Scientific name: *Sceliphron caementarium* (Drury, 1773), Common Name: Black and Yellow Mud Dauber	Mud daubers (*Sceliphron caementarium*), sometimes called dirt daubers or mud wasps, these are thin, 1-inch (25 mm) long black wasps with yellow legs, and long yellow-thread waists. They are not considered aggressive wasps, It is 15 to 17 mm long, dull black, with the bases of the antennae, collar, and scutellum, a portion of the metathorax, petiole, and portions of the legs bright yellow (Figure 9).	Adults of both sexes frequently drink flower nectar, but they stock their nests with spiders, which serve as food for their offspring. They prefer particular kinds of spiders primarily prey on relatively small, colorful spiders, such as crab spiders (and related groups), orb weavers and some jumping spiders. Mud daubers cram as many as two dozen small spiders into a nest cell.	Solitary female builds a nest out of moist mud containing several parallel cell rows. A paralyzed spider is stuffed into each cell and one egg deposited on each spider. The female then closes the cell opening with mud (Figure 12).	They are also the main predator of the black and brown widow spiders

Contd...

Table 8.1–Contd...

Sl.No.	Species Classification	External Features	Food Source	Nest Location	Ecological Importance
		Cuckoo wasp			
8.	Superfamily: Chrysidoidea, Family: Chrysididae, Subfamily: Chrysidinae, Genus: Stilbum, Specific name: *cyanurum* Scientific name: *Stilbum cyanurum* (Spinola, 1806)	Jewel wasp, gold wasp or emerald wasp, are the common names of cuckoo wasp. They are highly sculptured, with brilliantly metallic bodies and bright coloration. The head and thorax may be brightly metallic., antenna with 13 articles, abdominal venter convex, three well developed exposed terga. Ovipositor long and robust, tenthredinoid parasites. Members of this species assume a defensive posture by rolling up into a ball, completely covering the ventral half of the head and thorax with the abdomen. Do not sting; Forewing with a stigma, tegula covering bases of both forewing and hindwing (Figure 13)	Cuckoo wasp larvae are ectoparasites of bee and wasp larvae, parasitize "harmful" hosts, *i.e.* larvae or prepupae in nests prepared and often defended by adults (wasps and bees). Although some feed on walking stick eggs. They are parasites that lay their eggs in the nests of bees, wasps, and certain other insects, which are called their hosts.	The female wets a point of the dry mud of the nest with a drop coming from her mouth parts and then touches it with her ovipositor. The operation, repeated a few times, brings to the penetration of the muddy wall and to the deposition of an egg into the cocoon of the *Sceliphron.*	Predator of dead spiders, true bugs, aphids, or thrips.
9.	Superfamily: Vespoidea, Family: Pompilidae (spider-hunting wasps), Subfamily: Pepsinae, Tribe: Pepsini, Genus: *Hemipepsist* Common Name: Tarantula hawk wasp	Medium-sized robust (body length 65 mm) wasp; males smaller than corresponding females. Body is large, dark and wings are yellowish in colour. In females Head shape very variable, from very strongly constricted behind the eyes to extremely strongly swollen, when the vertex is often also swollen, Abdomen without a definite petiole. Usually long-legged; but almost always Extremely fine hairs, varying to extremely long, coarse and dense, most often black, hair (Figure 14)	Tarantula hawks get their name because the females hunt Tarantulas (Family Theraphosidae), huge, hole-dwelling spiders that are common in the desert. Tarantula wasps are "nectarivorous". Adult tarantula hawks feed on flower nectar or the juice of damaged fruits.	The female wasp excavates a nest with her well equipped front legs that have a row of stout spines which serve as efficient rakes.	

Contd...

Table 8.1–Contd...

Sl.No.	Species Classification	External Features	Food Source	Nest Location	Ecological Importance
			Parasitica		
10.	Suborder: Apocrita, Superfamily: Ichneumonoidea Latreille, 1802, Families: Braconidae	Braconidae is a family of parasitoid wasps and one of the richest families of insects. Between 50,000 and 150,000 species exist worldwide. Entire body is orange colored abdomen large and swollen (Figure 15). Body entirely orangeyellow, with head thorax abdomen and legs, ocelli and antennae black, two ovipositor tube with fine and sharp brown colored sting.	They are solitary insects, and most are parasitoids–the larvae feeding on or inside another insect which finally dies.	Being parasitoid this wasp species never constructs nest. The female finds a host and lays an egg on, near, or inside the host's body.	Many species are egg-larval parasitoids; hence they are often utilized as biological pest control agents, especially against aphids.

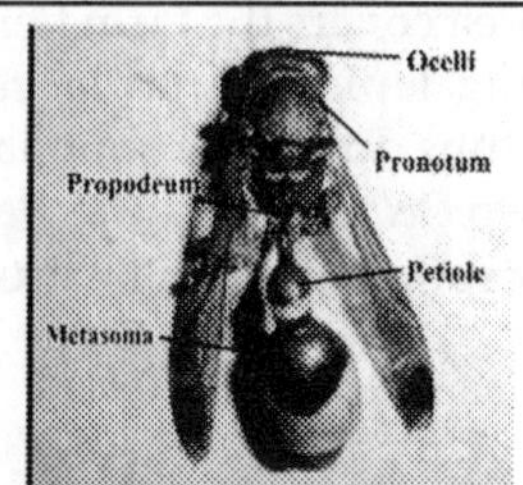

Fig.1 Adult *Ropalidia flavopicta*

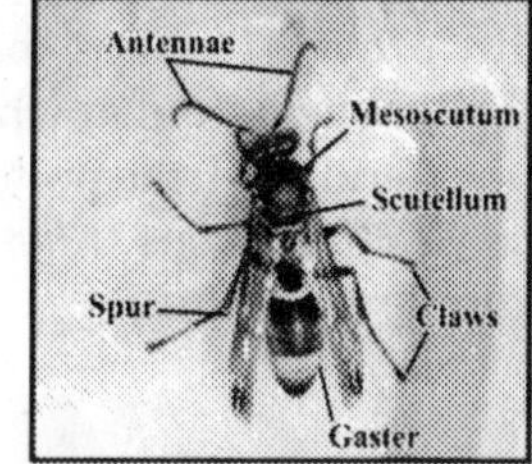

Fig. 2 Adult *Polistes humilis*

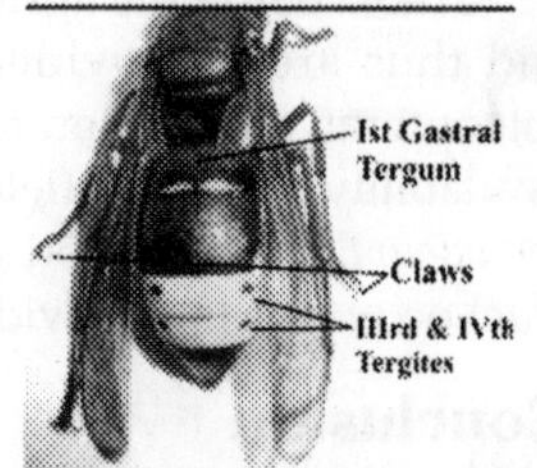

Fig. 3 Adult *Vespa orientalis*

Fig.4 Nest of *Ropalidia flvopicta*

Fig.5 Umbrella shaped nest of *polistes humilis*

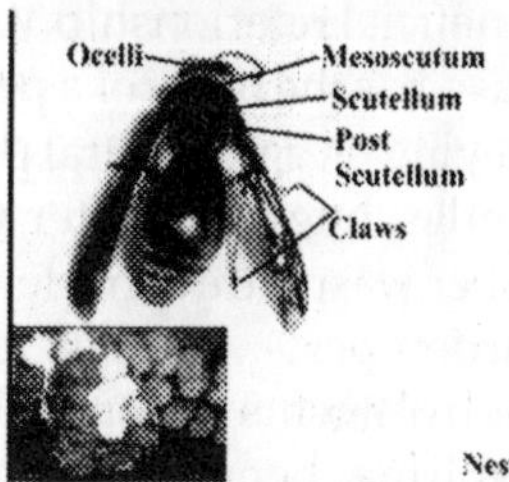

Fig.6 Adult *Rhynchium atrum with nest*

fig.7 Adult *Delta conoideum*

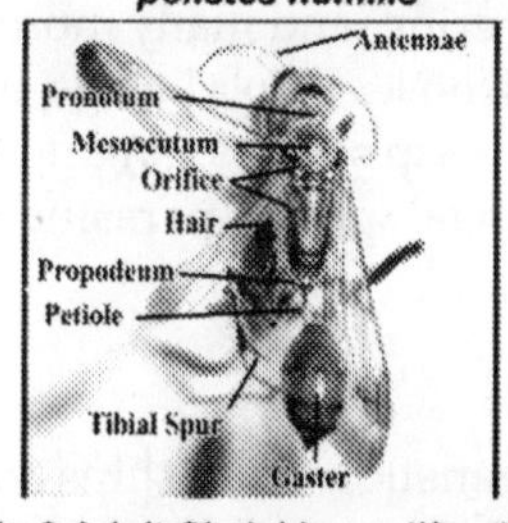

Fig.8 Adult Chalybion californicum

Fig.9 Adult *Schelifron caementarium*

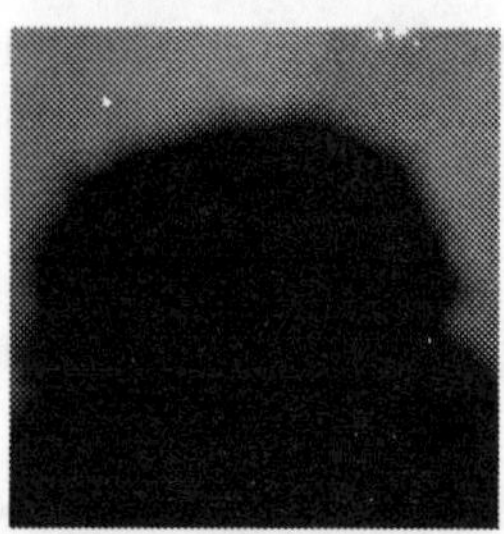
Fig.10 Nest of *Delta conoideam*

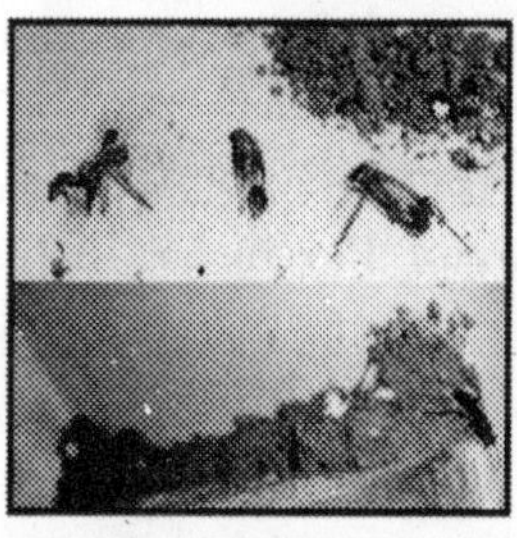
Fig.11 Nest of *Chalibion californicum*

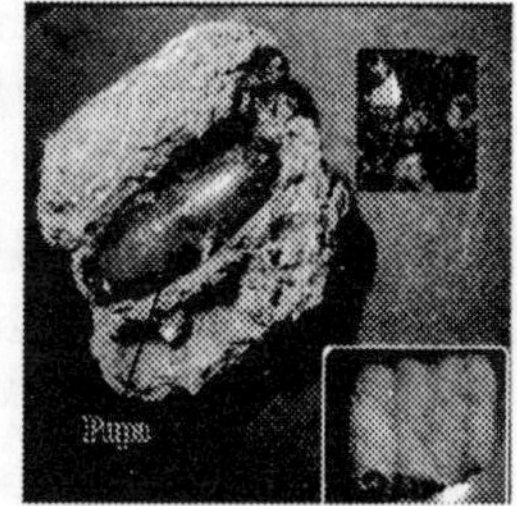

Fig. 12 Nest of *Schelifron caementrium*

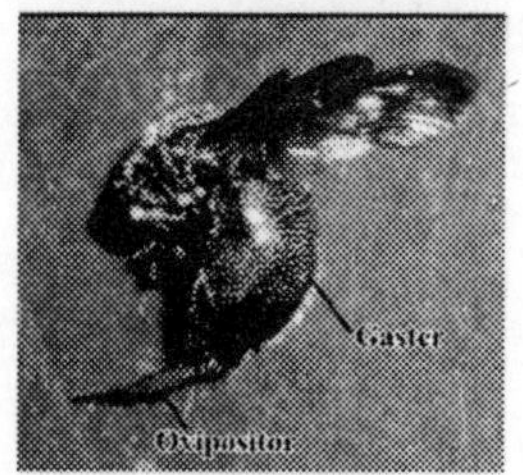

Fig. 13 Adultt *Stilbum cynarum*

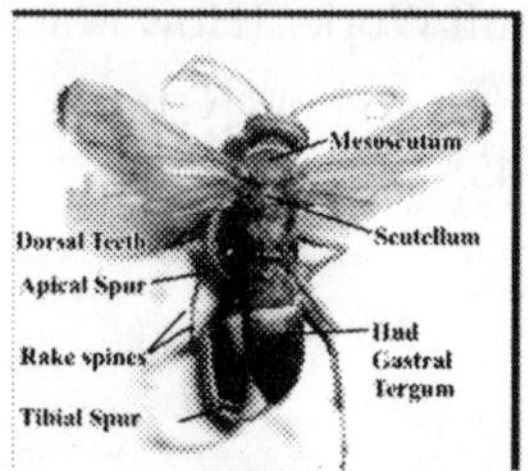

Fig.14 Adult *Hemipepsis* spp.

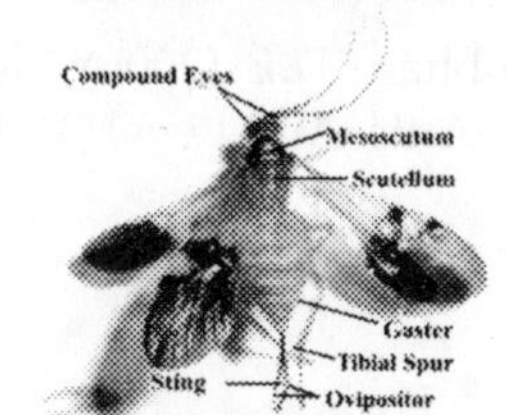

Fig.15 Adult *Braconidae spp.*

Akola being the agricultural city of Vidarbha, large area covers the farm fields and thus areas provided with preys of wasps including lepidopteran larvae, coleopteran and dipterans as well as variety of arachnids thus due to the abundant availability of muddy fields, water and food, the potter wasp *Delta esuriens* and *Delta conoideum*, Black mason wasps or mud wasps (*Rhyncium* sp.), metallic blue green Cuckoo wasps is also widely distributed here.

Conclusion

Wasps are highly important to ecosystems. Flowers and wasps share a mutually beneficial relationship.Wasps feed on flower nectar and play a role in pollination. A wasp has the body of a predator and not the forager. Wasps perform a vital service by preying on agricultural pests causing destruction of crop, such as caterpillars, aphids, beetles, bugs etc.Potter wasps are important in the natural control of caterpillars. Paper wasps are considered beneficial insects due to their predation on non-beneficial garden pests. Various ichneumons, Braconids are used successfully as biological control agents in controlling pests such as flies or beetles. These over all interesting, inspiring, beneficial and astonishingcharacteristics of wasps' leads to researches and investigations in various fields including: agricultural entomology, biotechnology, genetic engineering, and many more. In addition, the scope beyond imagination to workers and scientists. Akola being a unit of south east Asian continent shows high level diversity of wasp species ranging from social wasps(tribe or new world fauna), solitary or subsocial species, parasitic species, and more which are yet to be discovered.

References

Carpenter, J.M. (1987): Phylogenetic relationships and classification of the Vespinae (Hymenoptera: Vespidae). *Systematic Entomology* 12, 413–431.

Hermann, H. R. and Chao, J. T. (1984): Nesting biology defensive behavior of Mischocyttarus mexicanus cubicola (Vespidae: Polistinae). Psyche, 91:51-65.

Hermann, H. R. and Dirks, T. F. (1975): Biology of *Polistes annularis* (Hymenoptera: Vespidae). 1. Spring behavior. Psyche 85/1:97–108.

Hermann, H. R. and Chao, J. T. (1984): Nesting biology defensive behavior of *Mischocyttarus mexicanus* cubicola (Vespidae: Polistinae). Psyche, 91:51-65.

Kojima, J. (1999): Taxonomic notes on Australian *Ropalidia* Guérin-Méneville, 1831 (Hymenoptera: Vespidae, Polistinae).*Entomological Science* 2, 367–377.

Lamb, R. (1998-2011): How Stuff Works, (How Wasps Work), Inc.

Tembhare, D.B. (2006): Modern entomology, order–hymenoptera (Sawflies, Wasps, Ants, And Bees), 31:440-458

2013, Impact of Global Climate Change on Earth Ecosystems *Pages* ***75–82***
Editors: **D.R. Khanna, A.K. Chopra, Gagan Matta, Vikas Singh & Rakesh Bhutiani**
Published by: **BIOTECH BOOKS, NEW DELHI**

Chapter 9
Evaluation of Underground Drinking Water Contamination by Water Quality Index at Moradabad

Ashutosh Dixit[1], Navneet Kumar[2] and Nidhi Verma[3]
[1]Department of Chemistry, Singhania University, Pacheri Bari, Jhunjhunu, Rajasthan
[2]Department of Applied Sciences, College of Engineering, Teerthanker Mahaveer University, Moradabad – 244 001
[3]Department of Chemistry, NIMB University, Shobha Nagar, Jaipur, Rajasthan

Water quality indices for underground drinking water at Moradabad for four different sites have been calculated with the help of W.H.O. water quality standards and using the data of fifteen water quality physico-chemical parameters estimated following standard procedures and latest techniques of sampling and estimation. The drinking water of IM2 hand pumps at Moradabad is found to be severely contaminated at all the sites during the course of study. The present study suggests that people dependent on this water are prone to health hazards of contaminated drinking water. Some strict and effective measures for underground drinking water quality management are earnestly needed at Moradabad in the catchment area of study.

Keywords: *Drinking water contamination, Water quality index, Unit weight, Quality rating.*

Introduction

Water is the principal need of life on earth, and is an essential component for all forms of lives, from micro-organism to man. The unplanned urbanization and industrialization has resulted in over use of environment in particular of water resource. A kind of crises situation has made getting clean water a serious problem. Water is one of the most priceless gifts of nature. Today, water resources have been most exploited due to increasing population, industrialization, urbanization, increasing living standards and broad spheres of human activities. The level of pollution in the natural environment especially water contamination has been of great concern to the scientists, environmentalists and engineers because of its toxic nature and other adverse effects of human beings and other living creatures (Indirabai 2002, Karunakaran 2005). The problem of water contamination, water conservation and water quality management has assumed a very complex shape. Attention on this problem has become a need of hour because of far reaching impact on human health (Mahapatro 2004, Sinha 1995).

World oceans cover about three fourth of earth's surface. According to the UN estimates, the total amount of water on earth is about 1400 million cubic kilometers which is enough to cover the earth with a layer of 3000 meters depth. However the fresh water constitutes a very small proportion of this enormous quantity. About 2.7 per cent of the total water available on the earth is fresh water of which about 75.2 per cent lies frozen in Polar Regions and another 22.6 per cent is present as groundwater. The rest is available in lakes, rivers, atmosphere, moisture, soil and vegetation (Raju 2007, Mahadevsaswamy 2011). The crisis about water resources development and management thus arises because most of the water is not available for use and secondly it is characterized by its highly uneven spatial distribution. India's total renewable water resources are estimated at 1,907.8 km ^{3}/year. Its annual supply of usable and replanishable groundwater amounts to 350 billion cubic meters. Only 35 per cent of groundwater resources are being utilized. Groundwater supplies 40 per cent of water in India's irrigation canals. The Annual precipitation including snowfall is 4000 cu.km. The average annual water availability is 1869 cu.km and in 2001 per Capita water availability is 1820 cu.km. Total estimated utilizable water resources are 1123 cu.km. among which surface water resources are 690 cu.Km. and groundwater resources are 433 cu.km (Muduli 2010, Kavita 2010).

Water quality index (WQI) has been regarded as one of the most effective way to communicate the aquatic environment (Tiwari 1985, Sinha 1999). In a number of nationwide studies, it has been found to be quite useful in assessing the drinking water as well as river water quality physico-chemical parameters and their W.H.O. water quality standards are used for the calculation of water quality indices (WHO 1971, Sinha 2004, Pradhan 2001). The objective of the present study is to assess the underground drinking water quality at Moradabad during the course of study.

Moradabad is a 'B' class city of Uttar Pradesh having urban population more than 41 lakhs. Moradabad is situated at the bank of Ram Ganga River and its altitude from the sea level is about 670 feet. It is extended from Himalaya in north to Chambal river in south. Moradabad is at 28° 20', 29° 15' N and 78° 4', 79° E. District Bijnor and Nainital are in the north, Rampur is in the east, Ganga River is in the west and district Buduan is in the north of district Moradabad. Moradabad has seen rapid industrialization during last few decades. The major industries are – Brassware, Steelware, Paper mills, Sugar mills, Crushers and Dye factories etc. Most of these industries are playing their usual role in multiplying undergroundwater contamination. The catchment area of study is characterized by periodic occurrence of hot summers, moderate rains and cold winters. The maximum and minimum atmospheric temperatures are 48°C and 3°C respectively. The average rainfall varies between 800 to 1000 mm. The relative humidity is upto 90 per cent in monsoon and in drier part of the year it decreases to lees than 20 per cent.

Materials and Methods

India Mark II (IM2) hand pump water at four different sites at Moradabad district were collecting following standard procedures of sampling. The drinking water quality was assessed quantitatively following standard methods and techniques (APHA 1995, Merck 1974). The estimated physico-chemical parameters are pH value, turbidity, conductivity, alkalinity, total solids, total dissolved solids, total hardness, calcium, magnesium, dissolved oxygen, biological oxygen demand, chemical oxygen demand, free $CO_{2,}$ chloride and iron. All the chemical of anal R grade were used for experimental purposes. The specifications of instruments used are pH meter, Century CP 901 for pH value; Turbidity meter, Century nephelometer for turbidity; RI Conductivity meter for conductivity; Hach 2010 and Liberty AX sequential ICP-AES for iron concentrations.

A brief description of sites is as following:

I. Sri Grudwara, Chandranagar

This is IM2 hand pump water at a popular Grudwara in Chandranagar mohallah. This hand pump is extensively used by the residents of nearby locality for drinking as well as house-hold purposes. Water of this sites turns to light reddish yellow on standing for some time. The hand pump is claimed to be fixed at a depth of about 80 feet.

II. Singh Mandap, Chandranagar

This is IM2 hand pump water collected at Singh Mandap in Chandranagar mohallah. Water of this hand pump is used by ht e shopkeepers and residents of the locality for their daily usage. Water of this site is frequently used by the people using Singh Mandap for marriage and other purposes. A marked change in the colour of water of this site is noticed during the course of study.

III. Patelnagar

This is IM2 hand pump claimed to be fixed at a depth of about 90 feet at the beginning of densely populated Patelnagar mohallah. Area is characterized by the residents belonging to socially deprived class of society with lesser awareness for hygine. This hand pump water is extensively used for all the purposes as there is no water supply. This site is about 1.5 km west to site no. II. Water of this hand pump turns brownish yellow on standing.

IV Shiv Shakti Vatica, Loco-shed

This is IM2 hand pump water collected from Shiv Shakti Vatika of Shiv Shakti Mandir at Loco-Shed area. This site about 500 meter east to site no. II. Water of this hand pump is used by the residents of hydle colony and nearby locality for house-hold affairs. A sharp change of colour water to reddish brown on standing is noticed at this site. Some residents also reported heaviness in their stomach after intake of this water.

WQI of underground drinking water collected at four different sites at Moradabad were calculated using the methods proposed by Horton and modified by Tiwari and Mishra. (Horton 1965, Tiwari 1985) According to the role of various parameters on the basis of important and incidence on the overall quality of drinking water, the rating scales were fixed in terms of ideal values of different physic-chemical parameters. Even if, they are present, they might not be the ruling factor. Hence they were assigned zero values. Water quality indices of underground drinking water are calculated with the help of following equations:-

1. Quality rating, $Qn = 100\,[(V_n - V_i)/(V_s - V_i)]$
   here, V_n = actual amount present of nth parameter
   V_i = ideal value of this parameter
   $V_i = 0$, for all parameters except for pH and dissolved oxygen
   $V_i = 7.0$ mg/lit for pH; $V_i = 14.6$ mg/lit for dissolved oxygen,
   V_s = standard of this parameter
2. Unit weight, Wn for various parameters is inversely proportional to the recommended standard, Sn for the corresponding parameters.
   $Wn_1 = K/Sn_1$
   $$\sum_{n=1}^{n=15} W_n = 1$$
   considered here
3. Sub indices, $(SI)n = (Qn)^{Wn}$
4. The overall WQI is calculated by taking geometric mean of these sub indices.

$$W.Q.I. = \prod_{n=1}^{n=15} (SI)_n = \prod_{n=1}^{n=15} (Q_n)^{W_n}$$

OR

$$W.Q.I. = anti\log_{10}\left[\sum_{n=1}^{n=15} W_n \log_{10} Q_n\right]$$

To include the collective role of various physico-chemical parameters on the overall quality of drinking water, quality status is assigned on the basis of calculated values of water quality indices. The assumptions for extent of contamination or water quality status were: WQI < 50: fit for human consumption; WQI < 80: moderately contaminated; WQI > 80: excessively contaminated and WQI > 100: severely contaminated.

Results and Discussion

Parameter-wise WHO standards and their assigned unit weights (Wn) are listed in Table 9.1. Site-wise and parameter-wise estimated actual values (Vn) and calculated quality rating (Qn) are presented in Table 9.2. Finally calculated site-wise values of water quality index (WQI) are given in Table 9.3. Site-wise variations of water quality index values are presented graphically in Figure 9.1.

Table 9.1: Parameter-wise WHO Standards and their Assigned Unit Weights (Wn)

Sl.No.	*Parameter*	*W.H.O. Standard*	*Wn*
1.	pH	8.0	0.008821
2.	Turbidity (NTU)	5.0	0.014113
3.	Conductivity (µS/cm)	0.300	0.235217
4.	Alkalinity (mg/lit)	100.0	0.000705
5.	Total hardness (mg/lit)	100.0	0.000705
6.	Calcium (mg/lit)	100.0	0.000705
7.	Magnesium (mg/lit)	30.0	0.002352
8.	Chloride (mg/lit)	200.0	0.000353
9.	Total solids (mg/lit)	500.0	0.000141
10.	Total dissolved solids (mg/lit)	500.0	0.000141
11.	Dissolved oxygen (mg/lit)	5.0	0.014113
12.	Biological oxygen demand (mg/lit)	6.0	0.011761
13.	Chemical oxygen demand (mg/lit)	10.0	0.007056
14	Free CO_2 (mg/lit)	10.0	0.007056
15	Iron (mg/lit)	0.1	0.705650

A critical analysis of the values of WQI for IM2 hand pumps drinking water at Moradabad presented in Table 9.3 and its comparison with the standard assumption in meaningful facts regarding the undergroundwater contamination at Moradabad during the course of study in the catchment area of study.

The observed range of water quality index (WQI) is 195-259. The drinking water is found to be severely contaminated at all the sites of during course of study with WQI values more than 100. The IM2 underground drinking water is not found to be uncontaminated and fit for human consumption at all the sites during the course of study where the values of WQI are less than 50.

Table 9.2: Site-wise and Parameter-wise Estimated Actual Values (Vn) and Calculated Values of Quality Rating (Qn)

S.No.	*Patameter*	*Site No. I*		*Site No. II*		*Site No. III*		*Site No. IV*	
		Vn	*Qn*	*Vn*	*Qn*	*Vn*	*Qn*	*Vn*	*Qn*
1.	pH	7.23	23	7.14	14	7.35	35	7.25	25
2.	Turbidity (NTU)	27	540	30	600	26	520	28	560
3.	Conductivity (µS/cm)	.638	212.67	.762	254	.577	192.33	.648	216
4.	Alkalinity (mg/lit)	210	210	180	180	190	190	185	185
5.	Total hardness (mg/lit)	210	210	220	220	240	240	250	250
6.	Calcium (mg/lit)	165	165	170	170	185	185	190	190
7.	Magnesium (mg/lit)	45	150	50	166.67	55	183.33	60	200
8.	Chloride (mg/lit)	60	30	68	34	65	32.5	70	35
9.	Total solids (mg/lit)	701	140.2	775	155	660	132	720	144
10.	Total dissolved solids (mg/lit)	319	63.8	381	76.2	288	57.6	322	64.4
11.	Dissolved oxygen (mg/lit)	1.1	140.63	1.0	141.67	1.2	139.58	1.3	138.54
12.	Biological oxygen demand (mg/lit)	9.0	150	10.0	166.67	10.5	175	11.0	183.33
13.	Chemical oxygen demand (mg/lit)	42	420	41	410	42	420	40	400
14	Free CO_2 (mg/lit)	140	1400	155	1550	165	1650	175	1750
15	Iron (mg/lit)	.200	200	.190	190	.180	180	.260	260

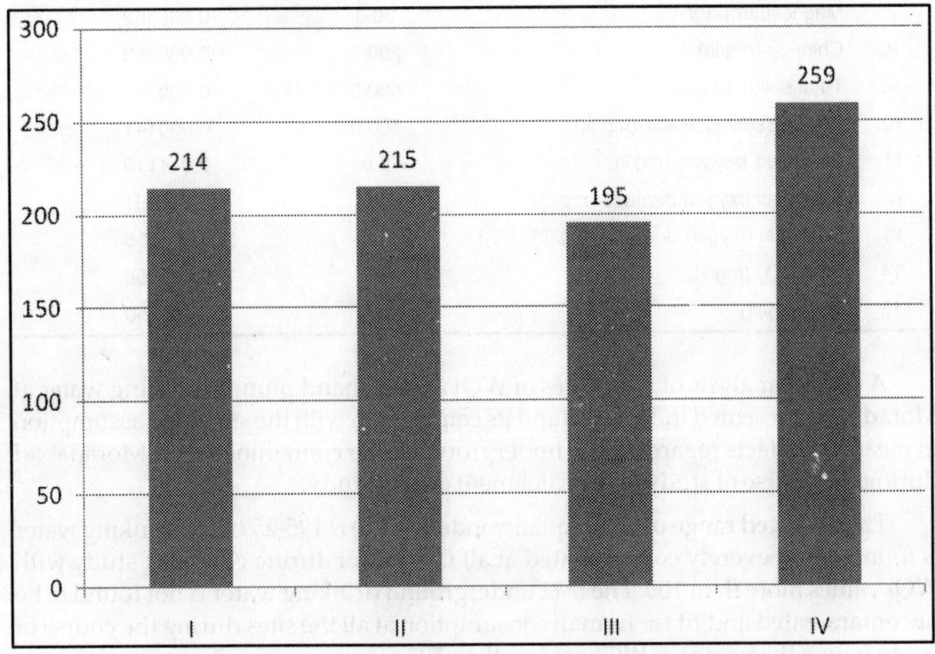

Figure 9.1: Site-wise Variations of Water Quality Index Values

Table 9.3: Site-wise Calculated Values of Water Quality Index (WQI)

Site No.	Name of Site	Water Quality Index
I	Shree Guru Duara, Chandra Nagar	214
II	Singh Mandap, Chandra Nagar	215
III	Patel Nagar	195
IV	Shatilok Shivalay Vatika, Loco Colony	259

Conclusion

On the basis of values of WQI, may be concluded that drinking water of IM2 hand pumps at Moradabad is severely contaminated and is unfit for human consumption and domestic usage at all the sites. In the other way it can also be said that underground drinking water at Moradabad is contaminated to some extent with reference to the water quality parameters studied. People dependent on this water might be facing the health hazards of contaminated drinking water.

The single line conclusion of the present study is that some strict and effective measures for drinking water quality management at Moradabad are earnestly needed and water quality assessment on the basis of water quality index is again proved to be an important tool.

Acknowledgement

The authors express their deep sense of gratitude to the Metal Handicraft Service Centre, Moradabad for providing the facility of inductively coupled plasma atomic emission spectrophotometer for iron concentration in the drinking water.

References

APHA, 1995. Standard method for examination of water and wastewater. 19th ed., AWWA, WPCF, Washington D.C.

Horton, R.K., 1965. An index number system for rating water quality. J. Water Poll. Cont. Fed. 37.300.

Indirabai, W.P.S. and George, S., 2002. Assessment of drinking water quality in selected areas of Tiruchurapalli town after flood. Poll. Res. 21(2): 209-214.

Karunakaran, K. *et al.*, 2005. A study on the physico-chemical characteristics of groundwater in salem corporation. India J. Env. Prot 25(6): 510-512.

Kavitha. R, and Elangovan. K. 2010. Groundwater quality characteristics at Erode district, Tamilnadu India. Inter. J. Environ. Sci. 1(2): 145-150.

Mahadevaswamy.G et.al, 2011. Groundwater quality studies in Nanjangud Taluk, Mysore District, Karnataka, India. Inter. J. Environ. Sci. 1(7): 1582-1591.

Mahapatro, T.R., 2004. Seasonal variation of fluoride in Rushikulya estuary, east cost India. Asian Jr. of Microbiol. Biotech. Env. Sc. 6(1): 37-40.

Merck, E., 1974. The testing of water. Darmstadt, Federal Republic of Germany.

Muduli Bipra Prasanna, Panda Chitta Ranjan, 2010. Physico-chemical properties of water collected from Dhamra estuary. Inter. J. Environ. Sci. 1(3): 334 – 342.

Pradhan, S.K., Patniak, D. and Rout, S.P., 2001. Groundwater quality index for groundwater around a phosphatic fertilizer plant. India J. Env. Prot. 21(4): 355-358.

Raju, N.J. 2007. Hydrogeochemical parameters for assessment of groundwater quality in the upper Gunjanaeru River basin, Cuddapah district, Andhra Pradesh, India. Environment and Geology, 18(3): 267-269.

Sinha, D.K. and Srivastva, A.K., 1995. Physico-chemical characteristics of river Sai at RaeBareli. India J. Env. Hlth. 37(3): 205-210.

Sinha, D.K., Saxena, S. and Saxena, R., 2004. Water quality index for Ram Ganga rivere at Moradabad. Poll. Res. 23(3): 527-531.

Sinha, A.P. and Ghose, S.K., 1999. Water quality index for river Yamuna. Poll. Res. 18: 435-439.

Tiwari, T.N. and Mishra, M., 1985. A preliminary assignment of water quality index of major Indian rivers. Indian J. Env. Prot. 5(4): 276-279.

W.H.O. 1971. International standards for drinking water. World Health Organization, Geneva.

2013, Impact of Global Climate Change on Earth Ecosystems *Pages 83–87*
Editors: **D.R. Khanna, A.K. Chopra, Gagan Matta, Vikas Singh & Rakesh Bhutiani**
Published by: **BIOTECH BOOKS, NEW DELHI**

Chapter 10

Study of Diversity of Phytoplanktons and Zooplanktons of Darna Dam in Nashik District (Maharashtra), India

V.R. Kakulte[1], D.F. Nikumbh[2] and B.J. Aher[3]
[1]Department of Zoology,
[2]Department of Botany,
[3]Department of Geography,
S.V.K.T. Arts, Science and Commerce College, Deolali Camp, Nashik, M.S.

A detailed investigation on the qualitative as well as quantitative plankton organisms from Darna dam was conducted. Twelve species of phytoplanktons were recorded. A number of phytoplankton increased in the month of December and decreased in the month of April eight species of zooplanktons were recorded. Zooplanktons showed quantitative variations. A no. of zooplanktons increased in the month of April and decreased in the month of November.

Keywords: Darna dam, Phytoplankton, Zooplankton.

Intoduction

The phytoplankton population in any aquatic system is biological wealth of water for fishes and constitute a vital link in the food chain, whereas the zooplankton is an important source of food for fish in the water body. It is evident that some fishes are typically active feeders on the plankton while others feed zooplankton throughout their life. So basically all the fishes are plankton feeders at some stages of their life cycle. The knowledge of plankton composition, abundance and distribution helps to evaluate their significance as fish food. For proper growth and maintenance of a water body, phytoplankton and zooplankton have their definite roles of any phytoplankton and zooplankton in a water body at micro level can be understood only when adequate data and the above two planktonic biomass are available.

The selected water body is rectangular, shallow and perennial one having concave basin and is used for the culture of fishes, especially major carps. The water body is getting eutrophicated fastly and blooming phytoplanktonic taxa, especially Chlorophyceae.

Material and Methods

The collection of phytoplankton was made by hauling two to three litres of water through the plankton net. Zooplankton were collected along with water. The collections were brought to the laboratory for further investigation. A sample of 200 ml was taken out. The zooplankton were first identified in living condition and then preserved in 5 per cent formalin.

Results and Discussion

Green algae were found present throughout the years of investigation but in lesser numbers. The maximum 50 to 80 per cent was recorded in December and minimum 25 to 35 per cent in April. No fixed pattern in the density of filamentous and nonfilamentous forms could be assessed. Since it was during December and January months, the nonfilamentous forms were in plenty and during other months of the year, the filamentous forms. The contribution of these algae towards total density of phytoplankton ranged between 35 to 80 per cent during different months of the year, while spirogyra, Hydrodictyon, Oedogonium, Cosmarium, Scenedesmus, were amongst the nonfilamentous ones. By order of frequency of occurance during the entire period of study, the genera encountered were *Spirogyra, Zygnema, Hydrodictyon, Chlamydomonas, Cosmarium* and *Oedogonium* sps. A total 12 species to 10 genera were collected during the period of investigation. They were enlisted along with seasonal changes in their density and abundance (Table 10.1).

Zooplankton of the investigation of water body were collected in the year Jan 2010 to December 2010. and their percentage was calculated in Table 10.2. Eight species of zooplankton were collected during different months of the year, highest value was recorded in the month of June. Their increasing trend was found to be from February to June and again in December their decreasing trend was recorded November to February. Thus 2 maxima in June and October and two minima in February and December were recorded. Eight species of zooplankton were found to be collected during the period of investigation. Major constituent were Crustacean, Hemiptera, Diptera. Among insects maggots larvae and wriggler larvae were found regular and abundant. Some insect larvae were irregular in shallow depth zones but were completely absent in the deep zones. The various zooplanktons were not equal

Table 10.1: Phytoplankton Population of Darna Dam during January 2010 to December 2010

Sl.No.	Name of the Phytoplankton sps.	Jan.	Feb.	Mar.	Apr.	May	Jun.	Jul.	Aug.	Sept.	Oct	Nov.	Dec.
1.	*Chaetophora* sps.	*	*	**	*	*	*	*	**	***	**	*	*
2.	*Chlamydomonas* sps.	***	*	*	*	*	*	**	*	**	*	*	*
3.	*Chlorella*	*	–	*	–	–	–	**	–	*	*	**	*
4.	*Closterium crosum*	**	*	**	*	*	*		*	****	***	*	–
5.	*C. cucumis*	*	*	*	–	*	*	**	–	*	–	*	–
6.	*Hydrodactyon* sps.	**	*	*	–	*	*	**	–	*	–	*	–
7.	*Oedogonium*	**	*	*	**	*	*		*	**	*	*	*
8.	*Pediastrum duplex*	*	*	*	**	–	–	**	–	*	**	*	–
9.	*Spirogyra* sps.	**	*	–	*	*	*	*	–	*	*	*	*
10.	*Ulothrix* sps.	**	**	–	**	–	–	–	**	*	**	*	*
11.	*Volvox* sps.	**	*	–	*	*	*	**	*	*	**	**	–
12.	*Zygnema* sps.	*	*	*	**	*	–	*	**	–	*	**	*

Note:- ****: Abundant (60 to 100 per cent); ***: Common(30 to 59 per cent); **: Frequent(15 to 29 per cent); *: Rare (1 to 14 per cent); –: Absent.

Table 10.2: Zooplankton Population of Darna Dam during January 2010 to December 2010

Sl.No.	Name of the Zooplanktons	Jan.	Feb.	Mar	Apr.	May	Jun	Jul.	Aug	Sept.	Oct.	Nov.	Dec
1.	Crustacea Daphnia	*	*	**	*	***	***	**	**	*	**	**	**
2.	Branchipus Hemiptera	*	*	*	**	**	**	**	*	**	*	*	*
3.	Cicada	*	–	*	**	**	**	**	–	*	*	**	*
4.	Ranatra	*	–	*	*	**	*	*	**	**	**	**	*
5.	*Eristalis* sps.	**	*	*	**	**	***	****	*	*	**	*	*
6.	*Psychoda* sps.	*	*	*	*	**	**	***	**	*	**	**	*
7.	Maggot larvae	*	**	***	**	**	**	**	–	*	**	*	*
8.	Wriggler larvae	–	**	***	**	**	**	**	–	*	**	–	–

Note:- ****: Abundant (60 to 100 per cent); ***: Common(30 to 59 per cent). **. Frequent(15 to 29 per cent); * Rare (1 to 14 per cent). – Absent in their abundance in the shallow zones or in the deep zones, but some were equal in both the zones in their abundance (Table 10.2).

During the investigation, presence of zooplankton was maximum in the month of May and the minimum in the month of February. In American lakes, maximum zooplankton was observed in April and minimum in September. The differences in the occurrence of peaks in zooplankton might be due to differences in nature of water bodies, differences in the composition of abiotic factors of water and the soil and variations in the productivity of different water bodies. Some workers correlated bottom community with the fish productivity and accordingly this water body is most suitable for the pisiculture.

References

EGGLETON, F.E. (1931): A limnological study of the profound bottom fauna of certain freshwater lakes Col. Mon., 1: 231-232.

JANA, B. B. and MANNA, A.K. (1975): Seasonal changes of benthic invertebrates in 2 tropical fish ponds, J. freshwater Biol. 7: 129-136.

MANDAL, B.K. and MOITRA, S.K. (1975): Studies on the bottom fauna of a freshwater pond at Burdwan. J. Inland fish. Soc. 8:34-38.

MICHAEL, R.G. (1969): Studies on the bottom fauna in a tropical fish water pond at Hydrobiologia, 31(1): 203-229.

PRASAD, B.B. and SINGH, R.B.(2003): Composition, abundance and distribution of Phytoplankton and Zoobenthos in a tropical water body. Nat. Env. Andpoll. Tech-2(3):255-258)

SHEEBA S. and RAMANUJAN, N.(2005): Qualitative and quantitative study of zooplankton in lithikkara river Res. 24 (1) -119-122.

SINGH, S.N.et. al. (1998): Hydrobiological studies of some eutrophic ponds of Rohtash, Bihar Env. and Ecology, 16(2): 457-462.

SHRIVASTAVA, V.K. (1956): Bottom organisms of a freshwater fish tank, Curr. Sci. 23: 158-159.

2013, Impact of Global Climate Change on Earth Ecosystems *Pages 89–94*
Editors: **D.R. Khanna, A.K. Chopra, Gagan Matta, Vikas Singh & Rakesh Bhutiani**
Published by: **BIOTECH BOOKS, NEW DELHI**

Chapter 11

Biodiversity and Rhizosphere Microfungi of Groundnut (*Arachis hypogea* L.) C.V. SB-11 with Special Reference of Fungicide Ceresan and Difolatan

R.S. Saler

P.G. Department of Botany K.T.H.M. College, Gangapur Road, Nashik – 2 Maharashtra

Use of fungicides has become an integral and economically essential part of agriculture when resistance against a pathogen in a variety fails. Many fungicides are directly introduced into agricultural lands for management of soil borne diseases. Some part of the fungicide compounds administered as spray or dusts ultimately reaches into the soil as run off or drift and thus influence the microbial balance of soil. There fore, the study of tolerance of pesticides by micro-organisms has got much practical importance in maintaining the survival of saprophytic micro flora in soil. Now- a-days, the necessity of such study is emphasized all over the world. In order to understand tolerance of Ceresan and Difolatan present study under taken. These fungicides are generally used in the management of groundnut leaf spot and rust diseases of groundnut. Ceresan and Difolatan were used and tolerance was studied by serial soil dilution plate count technique. Altogether 21 and 18 fungal species were recorded respectively in Ceresan and

Difolatan. Out of 21fungal species of which 14 were from the rhizosphere (R) where as 17 were from soil (S). Only *Aspergillus carbonarius* (R and S) showed resistance 80 mg/ml compound Ceresan from both rhizosphere and soil. A total of 18 species of fungi were recorded from Difolatan. Only *Aspergillus carbonarius* (R and S) showed tolerance up to 80 mg/ml. 10 mg/ml tolerance shown by *Mucor circinelloides* (R), *Rhizopus stolonifer* (R), *Aspergillus niger* (R), *Trichoderma viride* (R) while 30 mg/ml tolerance was shown by *Aspergillus fumigatus* (R), *Fusarium oxysporum* (S). Eleven fungal species did not tolerate 10 mg/ml of the fungicide including a potential pathogen *Rhizoctonia bataticola.*

Keywords: Tolerance, Rhizosphere (R) Soil (S), Fungi, Fungicide.

Introduction

Soil microorganisms play a very important role in maintaining soil fertility. However, concern has been expressed that agricultural chemicals may harm the soil micro flora (Martin,1950; Bollen,1979; Domsch,1964; Gangawane,1972; Subba Rao, 1977; Greaves, 1979; Saler and Gangawane, 1980; Gangawane and Saler, 1988; Saler and Gangawane, 1993, 1994; Gangawane *et al.*, 1995; Saler and Reena Chauhan, 2006) although the use of agricultural chemicals including pesticides has become an integral and economically essential part of agriculture. Anderson (1978) and Saler (1982) has reviewed the pesticide effect on soil microorganisms and it is clear that pesticides (herbicides, fungicides, insecticides) effect on soil microorganisms (bacteria, actinomycetes as well as fungi).The concept of rhizosphere effect where by the roots clearly bring out the beneficial effect of seed and soil mycoflora on growth of plant (Rangaswami, 1972). Therefore, tolerance of pesticides by these organisms is grate practical importance as far as their survival is concerned. Present communication reports the tolerance of Ceresan and Difolatan both non-target and target organisms in the rhizosphere of groundnut SB-11.

Materials and Methods

Groundnut (Arachis hypogea L.) C.V. SB-11 was used in this study. Seeds were sown in earthen were pots (6"x 6"x 6") using garden soil. They were observed for the germination after 15 days. Plants were collected to study the tolerance of Ceresan and Difolatan by rhizosphere micro fungi. Tolerance of rhizosphere micro fungi to Ceresan and Difolatan was studied by modified food poisoning soil dilution (FPSD) technique (Saler and Gangawane, 1980) using Waksman's acid agar medium. Equal volume of 2X above medium (served as food) and 2 X concentration of Ceresan and Difolatan (served as poison) along with 1ml of spore suspension from dilution flask (Served as soil dilution). Thus, the medium has the final concentration 10,30, 50 and 80 mg/ml. Media with single strength without Ceresan and Difolatan served as control. Observation for the number and types of micro fungi present on the plate were made from seven days on wards.

Results and Discussion

It was observed that quantitatively fungal populations were reduced in both rhizosphere and soil, as the concentration increased in the medium (Table 11.1).

Table 11.1: Number of Fungal Colonies (10^3/g oven dry soil) Tolerant to Different Concentrations (µg/ml) of Ceresan in the Rhizosphere and Soil of Groundnut-SB-11

	Control	*10*	*30*	*50*	*80*
Rhizosphere (R)	11.61	4.20	0.14	0.14	0.14
Soil (S)	3.15	1.35	0.11	0.11	0.11
R/S	3.60	3.11	1.28	1.28	1.28

There were about 8 and 2 fold reductions in the fungal population of rhizosphere and soil respectively. Statistically this negative correlation was non- significant for rhizosphere (r = –0.76) and soil (r = –0.77). This fungicide did not show its tolerance limit beyond 100mg/ml. Hence the concentrations were arranged below 100mg/ml in the medium. However fungal colonies appeared only 10 mg/ml except *Aspergillus carbonarius* which showed its final tolerance limit up to 80 mg/ml to this fungicide.

Table 11.2: Number of Fungal Species Tolerant to Different Concentrations (mg/ml) of Ceresan in the Rhizosphere(R) and Soil (S) of Groundnut- SB-11

Sl.No.	*Fungal Species*	*Control*		*100*		*500*		*1000*		*1500*	
		R	*S*	*R*	*S*	*R*	*S*	*R*	*S*	*R*	*S*
1.	*Absida corymbifera*	–	2	–	–	–	–	–	–	–	–
2.	*Actinomucor elegans*	1	–	–	–	–	–	–	–	–	–
3.	*Choanophora* sp.	1	–	–	–	–	–	–	–	–	–
4.	*Mucor circinelloides*	2	1	2	–	–	–	–	–	–	–
5.	*Rhizopus stolonifer*	4	2	2	–	–	–	–	–	–	–
6.	*Penicillium brefeldianum*	3	2	–	–	–	–	–	–	–	–
7.	*Aspergillus aculeatus*	3	2	–	2	–	–	–	–	–	–
8.	*A.carbonarus*	6	4	2	1	1	1	1	1	1	1
9.	*A.flavus*	4	2	4	–	–	–	–	–	–	–
10.	*A. fumigatus*	–	–	2	2	–	–	–	–	–	–
11.	*A. niger*	2	6	3	2	–	–	–	–	–	–
12.	*A. sclerotiorum*	–	3	–	–	–	–	–	–	–	–
13.	*A. terreus*	–	3	–	–	–	–	–	–	–	–
14.	*Aureobasidium pullulans*	3	1	–	–	–	–	–	–	–	–
15.	*Cladosporium oxysporum*	–	1	–	–	–	–	–	–	–	–
16.	*Fusarium oxysporum*	5	4	2	2	–	–	–	–	–	–
17.	*F. semitectum*	–	4	–	–	–	–	–	–	–	–
18.	*Penicillium funiculosum*	5	4	4	1	–	–	–	–	–	–
19.	*P. varians*	2	1	–	–	–	–	–	–	–	–
20.	*Rhizoctonia bataticola*	–	2	–	–	–	–	–	–	–	–
21.	*Trichoderma viride*	1	1	–	–	–	–	–	–	–	–

Qualitatively a total of 21 fungal species were recorded during this experiment (Table 11.2). Of these, 14 were from the rhizosphere where as 17 were from soil. Only Aspergillus carbonarius (R and S)showed tolerance up to 80 mg/ml. Species tolerated up to 10 mg/ml were: *Mucor circinelloides* (R), *Rhizopus stolonifer* (R), *Aspergillus aculeatus* (S), *A. A. flavus (R), A. fumigatus* (R and S), *A.niger* (R and S), *A. terreus* (R and S), Fusarium oxysporum(R and S), *Penicillium funiclosum* (R and S). Remaining 11 micro fungi did not tolerate 10 mg/ml, of the fungicide Ceresan including *Rhizoctonia bataticola* (R).

Table 11.3: Number of Fungal Colonies (10^3/g oven dry soil) Tolerant to Different Concentrations (µg/ml) of Difolatan in the Rhizosphere and Soil of Groundnut-SB-11

	Control	*10*	*30*	*50*	*80*
Rhizosphere (R)	11.61	0.72	0.29	0.14	0.14.
Soil (S)	3.15	0.45	0.11	0.11	0.11
R/S	3.60	1.61	2.57	1.28	1.28

Table 11.4: Number of Fungal Species Tolerant to Different Concentrations (mg/ml) of Difolatan in the Rhizosphere(R) and Soil (S) of Groundnut- SB-11

Sl.No.	*Fungal Species*	*Control*		*10*		*30*		*50*		*80*	
		R	*S*	*R*	*S*	*R*	*S*	*R*	*S*	*R*	*S*
1	Absida corymbifera	–	2	–	–	–	–	–	–	–	–
2	*Mucor circinelloides*	2	1	1	–	–	–	–	–	–	–
3	*Rhizopus stolonifer*	4	2	1	–	–	–	–	–	–	–
4	*Penicillium brefeldianum*	3	2	–	–	–	–	–	–	–	–
5	*Aspergillus aculeatus*	3	2	–	–	–	–	–	–	–	–
6	*A. carbonarius*	6	4	1	1	1	1	1	1	1	1
7	*A. flavus*	4	2	–	–	–	–	–	–	–	–
8	*A. fumigatus*	–	–	–	1	1	–	–	–	–	–
9	*A. niger*	2	6	1	–	–	–	–	–	–	–
10	*A. sclerotiorum*	–	3	–	–	–	–	–	–	–	–
11	*Aureobasidium pullulans*	3	1	–	–	–	–	–	–	–	–
12	*Cladosporium oxysporum*	–	1	–	–	–	–	–	–	–	–
13	*Fusarium oxysporum*	5	4	–	–	–	1	–	–	–	–
14	*F. semitectum*	–	4	–	–	–	–	–	–	–	–
15	*Penicillium funiculosum*	5	4	–	–	–	–	–	–	–	–
16	*P.varians*	2	1	–	–	–	–	–	–	–	–
17	*Rhizoctonia bataticola*	–	2	–	–	–	–	–	–	–	–
18	*Trichoderma viride*	1	1	2	–	–	–	–	–	–	–

In Difolatan final tolerant limit of fungal population in the rhizosphere and soil was 80mg/ml. There was negative correlation between the fungal population and fungicidal concentrations in the plate (rhizosphere r = –0.62; soil, r = –0.65). More reduction was seen at 10 mg/ml, whereas at 30,50,80 mg/ml population more or less same.Rhizosphere effect was more at 30 mg/ml, when compared to other concentration.

Qualitatively 18 fungal species were reported, only *Aspergillus carbonarius* (R and S) showed its tolerance up to 80 mg/ml. 10mg/ml tolerance was shown by: *Mucor circinelloides* (R), *Rhizopus stolonifer* (R), *A. niger* (R), *Trchoderma viride* (R), while 30 10mg/ml tolerance was shown by *Aspergillus fumigatus* (R), *Fusarium oxysporum* (S). Eleven micro fungi did not tolerate 10mg/ml of the fungicide including a potential pathogen *Rhizoctonia bataticola* (Table 11.4).

Acknowledgements

I am grateful to Dr. V. B. Gaikwad Principal and Prof D.R.Mahajan, Head, Dept.of Botany K.T.H.M.College, Gangapur Road, Nashik-2. for encourage and interest in the work.

References

Andeson,J.R.(1978) Pesticide effect on non-target soil microorganisms, Acad. Press. London, pp.313-533.

Bollen, G.J.(1979) Side effect of pesticides on microbial interaction, Acad.Press, London. Pp 451-481.

Domsch,K.H.(1964) Ann.Rev.Microbiol.15:69

Gangawane, L.V.(1972) Taxonomy and Physiology of Rhizosphere Mycoflora of Groundnut -SB-11. Ph.D Thesis, DR.B.A.M.U.Aurangabad.

Gangawane, L.V.R.S.Saler and U.D.Pingle (1995) Algalization in relation to nitrogen fixation in rice and rhizosphere mycoflora. Biologica Indica: 6(1 and 2) 11-16

Greaves, M.P. (1979) Proc. Assoc. Appli. Biol. 91:129.

Martin, J.P.(1950) Soil Sci. 69:107.

Rangaswami, G. (1972) Diseases of Crop Plants In India.Prentice Hall of India Pvt Ltd New Delhi, pp.456.

Saler,R.S.(1982) Investigation on the pesticidal tolerance by microorganisms in the Rhizosphere of groungnut. Ph.D Thesis. DR.B.A.M.U.Aurangabad

Saler,R.S. and L.V.Gangawane (1980) Tolerance of target and non –target spermosphere microfungi of groundnut to fungicides.Pesticides.14:22.-26

Gangawane,L.V. and R.S.Saler (1988): A comprative study of tolerance of fungicides *Rhizoctonia bataicola* and other microfungi in the rhihosphere of groundnut. Pesticies Ann.Rev.March 1988: 27-37.

Saler, R.S. and L.V. Gangawane (1993) Chemical Management of Plant Pathogens in Western India (L.V.Gangawane edtion) IPS (WZ) Publication, Aurangabad. pp.41-46.

Saler, R.S. and L.V. Gangawane (1994) In Vitro Tolerance of brassicol by rhizosphere micro fungi of groundnut (*Arachis hypogea* L). Indianbot.Reptr.13 (1+2): 41-46

Saler,R.S. and Reena Chauhan (2006) Tolerance of Actioncop by rhizosphere microfungi of groundnut C.V. SB-11. Bioifolet 3(3): 151-153.

Subba Rao N.S. (1977) Soil Microorganisms and Plant Growth.Oxford and IBH

2013, Impact of Global Climate Change on Earth Ecosystems *Pages 95–101*
Editors: **D.R. Khanna, A.K. Chopra, Gagan Matta, Vikas Singh & Rakesh Bhutiani**
Published by: **BIOTECH BOOKS, NEW DELHI**

Chapter 12
Environmental Conservation and Ancient Indian Scripts

Poonam Agarwal
Department of Botnay,
Govt. Girls P.G. College, Fatehpur, U.P.

Environmental science and ecology are disciplines of modern science but their origin can be seen long back in Vedic and ancient Sanskrit literature. The biological diversity and natural heritage of India were considered as Nature's gift protected by our ancestors. Our forefathers perceived God's presence around them through nature. Ancient Hindu scriptures such as Vedas, Upnishad, Puranas, Bhagwat Gita, Mahabharat, Ramayan *etc.* preach about valuing nature as nature sustains life and its destruction would inevitably endanger life on earth. Many hymns were written in the Vedas praising the mountains, rivers, forests, animals and sun and considered worthy of worship.

Vedas has depicted the close relationship between man, environment and religion. Religion was probably used in ancient India as a tool to protect environment and natural resources. The ancient Indian scripts contain the earliest message for conservation of environment and ecological balance. In general practice, there are many flora and fauna which are sacred because they all are directly or indirectly related to different God/Goddesses. The greatest honour given to animals was their elevation as the vehicles of the Gods/Goddesses. Similarly, many plants/trees were regarded as sacred because of their medicinal/ aesthetic/natural qualities as well as their close proximity to a particular diety and Hindu worship them regularly even today. Hindu religion scripts and mythology had left various religious beliefs for us towards nature and it was a very constructive device for conservation of biodiversity not only during olden times but at present also. The main reason for associating plants and animals as a part of religious rites or beliefs was probably for conservation of environment.

Keywods: Environment, Conservation, Natural resources, Religion, Ancient Scripts.

Introduction

Environmental degradation, one of the major problems the world is facing today. The problem of environmental pollution is not new but it is as old as the evolution of man on earth. But man's ambition for limitless comfort has led him towards the exploitation of nature's wealth and resources indiscriminately and rapidly. Man's greed for resources and his desires to conquer nature has placed him in front of the environment. The demand for new technological society has created a stress on environment. Therefore, conservation of environment has become now need of today.

It is generally stated that need of environmental conservation started only at the end of the 20th century after the Stockholm Conference in 1972. But it is not true as the environmental protection and conservation started in India from Vedic Period. Vedas has depicted the close relationship between man, nature and religion. Man from the ancient time had regarded environment as very important and that was why they worshipped the trees, animals, mountains and rivers etc. Religion was probably used in ancient India as a tool to protect nature and natural resources and several instances of worshipping the trees have reported from different periods of the country. Hindu religious scripts, mythology, and rituals have attempted to drive home the importance of preserving nature by deifying it through the centuries. Bhagavad-Gita (9.26) states:

Patram Pushpam phalam toyam, yo mey bhaktya prayataatmanha

Tadaham bhakt Yupahrutam asnaami prayataatmanaha

I accept a leaf, flower, fruit or water or whatever is offered with devotion

Hindu religion is often coined as a "environment friendly" religion. The Mahabharata, Ramayan, Vedas, Upnishada, Puranas and Bhagwat Gita contain the earliest message for preservation of environment and ecological balance.

Traditional Views of Environment

Vedic views revolve around the concept of nature and life. It describes the dynamics of five elements (panch tatwa) – earth, sky, wind, water and fire(sun). All the entities in the world are dependent upon these five elements.

Environmental science and ecology are disciplines of modern science but their origin can be seen long back in Vedic and ancient Sanskrit literature. Our ancient forefathers perceived God's presence around them through nature. They felt that they must live in harmony with all of God's creation, respect and revere all of nature and divine forces. They identified the divine forces as air, water, earth, fire and sky. Many hymns were written in the Vedas praising the elements, rivers, forests, animals and sun and considering worthy of worship.

The Holy Scriptures Bhagvatam (Vol. 2, chapter1, and verses 32-33) says- "*The air is his breath, trees are the hairs of his body, the ocean his waist, the hills and mountains are his bones, the rivers are the veins of the cosmic Being, His movements are the passing of ages.*" Even to this day, the significance of all these manifestations of God is understood, so they are worshipped by Hindus. Hindu mythology describes river Ganges as originating from the top of Lord Shiva's head in Himalayas.

Vedas and Upanishad mention that the God and Goddesses favour different biological resources. Yajurveda emphasized that the relationship with nature and animals should not be that to misuse the resources and therefore, there was a check on excess utilization of resources. Similarly, Rig-Veda is celebration of nature. It highlighted the potentialities of nature in controlling the climate increasing fertility. Rig-Veda mentions about forest goddesses and healing properties of plants. It also instructs that forest should not be destroyed. Atharvaveda considered trees as abode of various Gods and Goddesses. Also the Skand Purana states Tulsi and Bilva Tree as abode of Goddess Laxmi and Parvati respectively. Banyan tree is considered to be very sacred and worshipped as the abode of Trimurti's reside.

" *Parvati Bilva Vrikshastham Laxmi Cha Tulsi Gatam.*"

Atharvaveda also mentions certain names of Oushadhis with their values. Later this information became important source of Ayurveda. In general practice, there are many flora and fauna which are sacred among the followers of Vedas because they are directly or indirectly related with different Gods. Animals were revered too. Kamadhenu was the wish-fulfilling cow, whose offspring are all the cattle on earth. Krishna even lifts Mount Govardhana to save cattle from Indra's wrath, a recurring theme in Indian art. The festival of Nagpanchmi –snake worship is celebrated as a thanks giving after the harvest season. But the greatest honour given to animals was their elevation as the vehicles of the Gods/Goddesses (Table 12.1).

Table 12.1: A General List of Animals Favoured by God/Goddess

Animal	*Favoured by God/Goddess*
Owl	Goddess of wealth (Laxmi)
Bull	Lord of animal (Pashupatinath),Lord Shiva
Lion and Tiger	Goddess of power (Durga, Kali etc,)
Horse	Lord Sun
Serpent	Lord Shiva
Monkey	Lord Hanuman
Dog	Lord Bhairab
Rat	Lord Ganesh
Swan	Goddess of knowledge(Saraswati)
Eagle (Garun)	Lord Vishnu
Cow	Lord Krishna

Similarly, there are many sacred plants and Hindus worship them regularly *e.g. Oscimum sanctum* (Tulsi), *Eleocarpus sphaericus* (Rudrakshya), *Ficus bengalensis* (Bar),

Ficus religiosa (Pipal), Prosopis cineraria(Shami) etc. Some of these herbs were believed to cure every human diseases from head to toe. A few of these plants were so common in our daily life and their therapeutic value was considered so great that they became a part of religion and worship by Hindus. *e.g.* Oscimum sanctum (Tulsi),Phyllanthus emblica (Amla) etc. In Rig-Veda, ' Soma' (*Ficus benjamina*) is mentioned as king of plants. Sometimes threads are tied around certain trees to seek a boon. Nyctanthus arbor tristis (Parijat/Harsingar) is another divine tree believed to be introduced by Lord Krishna from the Heaven. There are hundreds of medicinal plants which are in use from the Vedic periods till today. Oushadhi Sukta of Rig-Veda addresses to plants and vegetation as mother,' *O Mother ! Hundreds of your birth places and thousands of your shoots*'. Several Indian trees and shrubs were regarded as sacred because of their medicinal/aesthetic/natural qualities as well as their proximity to a particular deity. Tree symbolized various attributes of Gods to Rig-Veda Seers. Association of trees, plants and animals with various Gods and Goddesses at that time testified reverential attitudes of people towards the environment (Table 12.2).

Table 12.2: A General List of Plants/Trees Favoured by God/Goddesses

Tree/Plant	*Favoured by God/Goddesses*
Neem	Sitla, Mansa
Banyan	Sheshnag
Tulsi	Lakshmi
Vishnubela	Shiva
Ashok	Buddha, Indra
Kadamba	Krishna
Mango	Lakshmi
Pipal	Vishnu, Krishna
Bel	Parvati

According to Purana,"One tree is equal to ten sons". Significance of plants and trees to human life is further exemplified in Varahpurana which advocates regular plantation as a means to achieve heaven. To be particular, it says:

Ashwathmekam piumindamekam nuogredhmekam das pushpa jati

Dwe dwe tatha darim matulunge panch amra ropi narakam na yati

i.e. "One who plants one pepal, one neem, one bar, ten flowering plants or creepers, two pomegranates, two oranges and five mango trees will not go to hell".

According to Vedic traditions, every village will attain wholeness only when the forest is present. Some of these are however equivalent to the 'Protected Areas' and 'Production Forests' of today. In Charak Samhita, destruction of forest is taken as destruction of the state and reforestation, an act of rebuilding the state and advancing its welfare. Every village must have a cluster of five great trees- Panchvati, symbolizing the five primary elements; earth, water, fire, air and ether- the totality of every thing.

Concept of Environmental Conservation

Sustainability was in thought process of the early Indians as evident from the teaching of Vedas. Hymn from *Atharveda* (12:1:35) says – *"Whatever I dig out from you, O Earth! May that have quick regeneration again; may we not damage thy vital habitat and heart"*. Vedas believe that the earth is an object of worship because she is the basis of life. Man is part of earth like other creatures or elements which are equally dwelling. Therefore, ecology was a sacred science for the Vedic man. Mahatma Gandhi observed: *"I bow my head in reverence to our ancestors for their sense of the beautiful in nature and for their foresight in investing beautiful manifestations of nature with a religious significance."*

Pollution control is becoming a challenge to the scientists and failure of the Copenhagen Summit speaks eloquently the graveness of the situation. The source of present pollution is none other than our anti-ecological modern science! That is why, the present day since is finding it difficult to solve this problem effectively. Ancient Indian Science was always Nature Friendly and hence, this problem was not so rampant at that time. The Vedas attach great importance to environmental protection and purity. They insist on safeguarding the habitation, proper afforestation and non pollution.

Lord Krishna spread the message that nature needs to be preserved. Bhagvata Gita 3:12 says that – *"For so sustained by sacrifice, Gods will give you the food of your desire. Whoso enjoys their gift, yet gives nothing, is a thief, no more, no less."*

The Vedas stress the need for protection and development of forests. The Vedas assert that the plants and trees are verily the treasures for generations. Mythology also has been useful in cultivating certain plants *e.g.* Tulsi, a highly valued medicinal plant is grown in every household and ritually watered daily even today. This is clearly a case of religion and culture being used to protect, conserve and/or produce resources from human sustenance. The students of Vedic period lived in a natural environment (Gurukul), the students as well as their gurus were concerned about preservation of environment. They protected trees and worshipped them as Vriksha Devta (tree god), the forest covers as Van Devta (forest god) and the rivers as sources of delicious life-giving water. The ancient people cared for wildlife too. Terms and titles such as Nag Devta (snake god), Kamdhenu (the cow that fulfills your desires) and Kalpavriksha (the wish –fulfilling tree) symbolized the benefits that accrued to human beings from nature and their respect for wildlife. Thus pantheism or animism, by whatever name we may call it, eventually pointed to ecological balance and conservation of nature. The ecological imbalance caused by the criminal acts of the so called 'civilized man' has resulted in a disastrous threat, not only to the human survival, but also to life as a whole on the earth. Concept of *Yagya* seems to be a viable cure for the ecological imbalance. Experimental studies have shown that *Yagya* or *agnihotra* creates a pure, hygienic, nutritional and healing atmosphere. Researchers from the field of microbiology have observed that the medicinal fumes emanating from the process of *agnihotra* are bacteriostatic in nature, *i.e.* they eradicate bacteria and micro-organisms, which are the root causes of illness and diseases. The example of tragic incident of 3rd Dec.1984, MIC gas leakage of Union Carbide is worth mentioning. Hundreds of people died and many hospitalized but two families of

Bhopal came out unscathed because they were regularly performing 'Havan'. Thus, 'agnihotra' is antidote to air pollution. It helps in increasing crop fields in protecting plants from diseases as well as providing disease free pure and energized environment for all offering peace and happiness of mind.

India is a land of rites and rituals. Almost all major religions of the world are represented in India. Besides Hinduism, all other religions also realized the proximity of mankind with nature. For example- Basic tents of Buddhism are simplicity and non-violence. Both these principles of Buddhism are of great importance in the conservation and protection of natural environment. The principle of simplicity teaches us that man should not overexploit the natural resources *i.e.* it is based on sustainability which is also the crying need of the present time. Non –violence shows love for flora and fauna, and respect for rivers, forests, mountains, the sun and moon. Similarly King Ashok wanted the non-violence to be the cultural heritage of the people. Therefore, punishment was prescribed for killing animals. Jainism is also based on the principle which is in close harmony with nature and propagates respect for nature's creation and their protection. In Islam also, there is close harmony between man and nature. The Quran is sensitive to the cutting of tree. The message of Pope Paul VI to United Nations Conference held at Stockholm in June 1972, states that there is a close link between Christianity and environment and thrust for sustainable development. Sikh religion is comparatively of recent origin. Guru Granth Sahib Ji also emphasizes that human being are composed of five basic elements of nature. Thus close relationship between nature and mankind has been recognized.

Therefore, the cultural and religious heritage of India shows deep concerns for the protection and preservation of the environment. All these religions realized the proximity of mankind with nature. Thus, the nature has been directly interconnected with religion and religion had a direct effect on the conservation and protection of the environment.

Practical Example from India

History of recent past tells us it was not only Kings or rulers of India who showed concern for nature but even common men were deeply involved in strong environmental practices. For instance, Bishnois of Rajasthan, a sect founded towards the end of 15th century actually involved themselves in the respect for all living things and their protection. It is believed that 360 of them mostly women and children sacrificed their lives by hugging the trees to save them from axe men of Maharaja of Jodhpur in 1730. Chipko Movements active since 1973, is one of the most successful conservation movements in India by womenfolk of Garhwal in Uttarakhand.

Conclusion

Vedic people desired to live a life of hundred years and this wish can be fulfilled only when environment will be unpolluted, clean and peaceful. An India's relation with nature differs from that of western men. In the West, man has separated himself from nature; he believes and used it to serve his own purpose. Hindu unites himself with nature. From nature he came, to nature he returns as ashes. Modern scientists also feel proud of our ancestors for their knowledge and views about the environment. Vedic message is clear that environment belongs to all living beings, so it needs protection by all, for welfare of all. Indian religious literature is replete with ideas of forest conservation, utilization and regeneration. Our ancestors had left various religious beliefs for us towards nature and it was a very constructive device for conservation of environment and biodiversity not only during their time but at present also. Therefore, main reason for associating plants and animals with religious rites was probably for conservation of biodiversity.

In recent years, we seem to have lost touch with our glorious tradition and wisdom of protecting nature. Environmental protection and sustainable development are concerns of such nature that permeate the entire humanity. In effect, environmental conservation is embedded in the Hindu consciousness: being a good Hindu means being a champion of the environment. Mahatma Gandhi has delivered inspirational message for environmental movement. He has said *"The country's development has to be in harmony with nature. The earth has resources to meet everybody's needs but not anybody's greed".*

From the foregoing account, it is evident that Indian rituals have been contributing a lot in protection and conservation of environment in India since Vedic Period. Hence, it is need of hour to go for these rituals and traditions in order to keep up the balance of nature. The ancient Indian message is that the environment belongs to all living beings, so it needs protection and conservation by all, for welfare of all.

References

Athraveda 12.1.35.

Asif Ahmad. Feb. 2011. Indian Environmental Concerns: Protection of Indian Religious and Cultural Heritage. Peace and collaborative Development network.

Bhagvatam, vol.2,chapter 1,verses32-33.

Bharat Budholai. Environmental Protection Laws in the British Era. Legal service India.com.

Giriraj Shah. Glimpses Of Indian Culture, p 106.

Jeetem shardah shatam, Atharveda,19.67.1.

Kumar,B.M. 2008. Forestry in Ancient India: Some Literary Evidences on Productive and

Protective Aspects. Asian Agri-History, vol. 12, no. 4, pp. 299-306.

Manisha Tyagi. Feb. 2011.Environmental wisdom in Ancient India. Variorum multidisciplinary e- research journal, vol. ii, issue iii.

Ranjan Malvika. 2010. Living in Harmony with Nature' message of Ancient Hindu Spritual Texts. Educational Quest,vol.1; issue 1.

Shatam vo amb dhamani sahastramut vo ruh. Rigveda, 10.97.2.

Surendra R. Devkota. July 2003. Veda and Ecological Economics: Ray of strong Sustainability; The Nepal Digest, year 14, vol.vii, issue 1.

Vanani nah prajahitani. Rigveda, 8.1.1

Vedic way to Beat Pollution. 1985. The Hindu- English daily, 4 May.

2013, Impact of Global Climate Change on Earth Ecosystems *Pages 103–114*
Editors: **D.R. Khanna, A.K. Chopra, Gagan Matta, Vikas Singh & Rakesh Bhutiani**
Published by: **BIOTECH BOOKS, NEW DELHI**

Chapter 13

Chemotaxonomy of Wild *Origanum vulgare* L. from Kumaun Himalayas

Shalini Singh, Geeta Tewari, Chitra Pande and Charu Singh

Department of Chemistry, Kumaun University, Nainital – 263 002 Uttarakhand

The herbal parts of *Origanum* species have been traditionally used by local people as folk medicine. The essential oil composition of aerial parts of *Origanum vulgare* L. (family: Lamiaceae) collected from ten wild locations, in Central Himalayas, India was analyzed by GC and GC/MS.

The cluster analysis confirmed the presence of four chemotypes with respect to the content of p-cymene/γ-terpinene/thymol/carvacrol; thymol/caryophyllene oxide/aliphatic hydrocarbons; linalool/thymol/carvacrol/β-caryophyllene/germacrene D and linalool/bornyl acetate/β-caryophyllene/germacrene D/bicyclogermacrene/elemol. On the basis of biosynthetic pathway these four chemotypes were further classified into two groups; phenylpropanoid group belonging to shikimic acid pathway whereas monoterpenoid group formed by the mevalonic acid pathway. The statistical significant correlation with in the major constituents of *O. vulgare* showed that p-cymene is positively correlated with γ-terpinene and carvacrol. γ-Terpinene was found to be positively correlated with carvacrol. Linalool is positively correlated with β-caryophyllene while negatively correlated with thymol and aliphatic hydrocarbons. Bornyl acetate in the plant essential oil was positively correlated with β-caryophyllene, germacrene D, β-bisabolene, elemol, and negatively correlated with thymol. Thymol was negatively correlated with β-caryophyllene, germacrene D and β-bisabolene. β-Caryophyllene was positively correlated with germacrene D and β-bisabolene. Germacrene D showed positive correlation with β-bisabolene and elemol. β-bisabolene was positively correlated with elemol while caryophyllene oxide showed positive correlation with aliphatic hydrocarbons.

Keywords: *Origanum vulgare L., Lamiaceae, Chemosystematic study, Cluster analysis, Thymol, Bornyl acetate.*

Introduction

Origanum vulgare L. commonly known as Himalayan marjoram in India belongs to family Labiateae (Lamiaceae). It is an erect perennial aromatic herb, with small pale, pink flowers crowded in to a branched domed inflorescence. This plant is 20 - 80 cm high with ovate entire and staked leaves 1-4 cm. It is commonly found in temperate regions from 7000 to 12000 ft. It is also distributed in Pakistan, Bhutan and temperate Eurasia at the height of 1500-3600 m (Polunin and Stainton, 1984), usually in chalky soil (Ietswaart, 1980). *O. vulgare* L. is used for perfumery, cosmetic preparations and medicines. It is also used in the production of scented grape wines (Bodrug, 1982).

The essential oil composition of *O. vulgare* L. has been reported from many countries (Chalchat and Pasquier, 1998; Russo, *et al.*, 1998; D'antuono *et al.*, 2000). *Origanum vulgare* L. growing in eight different localities of Vilnius district (Lithuania) has two chemotypes: ß-ocimene type and germacrene D type (Mockute *et al.*, 2001). Carvacrol has been reported as the major component of the *O. vulgare* ssp. *hirtum* (Baser *et al.*, 1994; Veres *et al.*, 2003). Three chemotypes of *O. vulgare* L. namely carvacrol/thymol, thymol/α-terpeniol and linalyl acetate/linalool growing wild in Campania (Southern Italy) have been reported by De Martino *et al.* (2009). D'antuono *et al.* (2000) reported three chemotypes of *O. vulgare*: carvacrol/thymol type, (*E*)-caryophyllene, γ-muurolene and high linalool type and β-bourbonene, (*E*)-caryophyllene, γ-muurolene, germacrene-D-4-ol and caryophyllene oxide type, collected from Northern Italy. Considering huge genetic diversity, the information on *O. vulgare* from India is scanty (Kaul *et al.*, 1996; Pande and Mathela, 2000; Bisht, *et al.*, 2009).

The essential oil composition of medicinal and aromatic plants is not constant but varies quantitatively and qualitatively. Essential oil quality depends upon different environmental factors like nature of soil, climatic conditions like light, altitude, moisture, growing and harvesting time etc. (Schearer, 1984). To best of our knowledge no work based on the chemosystematics of *Origanum vulgare* from different regions of Uttarakhand has been undertaken. Therefore, the objective of the present investigation is to explore the chemosystematics of this important genus in Uttarakhand, India.

Materials and Methods

Collection of Plant Material and Soil Samples

Fresh plant material of *O. vulgare* L. along with its soil samples (0-20 cm) were collected in September to November, 2009 from ten different locations *viz.* Dhoulchina (29°37′N, 79°40′E), Champawat (29°36′N: 79°30′E), Dharchula (29°51′00″N: 80°31′60″E), Munsiyari (30°04′37"N: 80°23′04"E), Ramgarh (29°23′N: 79°30′E), Kilbury (29°23′N: 79°30′E), Mukteshwar (29°28′N: 79°39′E), Mussoorie (30° 27′ N: 78° 06′ E), Nainital (29°23′N: 79°30′E) and Rushi village (29°23′N: 79°30′E) in Kumaon Himalaya (Uttarakhand, India). *O. vulgare* L. plants were in full blooming stage. The botanical identification of the specimen was done at Botany Department, Kumaon University, Nainital (Voucher no. -2036).

Isolation of Essential Oil

The fresh plant material was steam distilled for 3 h using a copper still fitted with spiral glass condensers and extracted with n-hexane and dichloromethane. The organic phase was dried over anhydrous sodium sulfate and the solvent removed by distillation using a thin film rotary vacuum evaporator at 25°-30°C. The oil yield was 0.3-0.4 per cent (v/w).

Analysis of the Essential Oil

The oil was analyzed by using a Perkin Elmer Autosystem XL GC. The column temperature was programmed from 70°-290 ° at 3°C/min using hydrogen as the carrier gas at 10 psi column head pressure. The injector temperature was 280°C and detector (FID) temperature was 290°C. The GC/MS used was Autosystem XL GC fitted with Equity-5 (60m x 0.32mm i.d., film thickness 0.25µm) fused silica capillary column coupled with Perkin Elmer Turbomass fitted with Equity-5 capillary column whose dimensions matched the column used with the GC. The column temperature ranged from 70°-300°C at 3°C/min using helium as carrier gas. The injector temperature was 220°C, and the injection volume 0.05 L in *n*-hexane, with a split ratio of 1:50. MS were taken at 70 eV with a mass range of 40-450 amu.

Identification of the Components

Sample components were identified by matching their mass spectra with those in WILEY and NIST, MS library search and by comparing with literature and GC retention indices (RI) (Adams, 2007)

Statistical Analysis

Ward's hierarchical clustering analysis of major constituents of essential oils was conducted in order to discriminate chemotypes. Cluster analyses were performed using SPSS 16.0 and the resulting dendrograms were illustrated. Experimental data were processed using Microsoft Excel XP. Correlation coefficients were calculated among the major constituents of oil. Significance level of correlation coefficient was checked on probability level of $pd \leq 0.05$ and $pd \leq 0.01$.

Results and Discussion

Chemosystematics of *O. vulgare* L.

The comparative results of *O. vulgare* L. collected from ten different locations shows difference in the chemical constituents of the essential oil allowing them to be characterized into four distinct chemotypes (Table 13.1). The oil of *O. vulgare* L. collected from Dhoulchina and Champawat (chemotype I) showed p-cymene (6.7-9.8 per cent), g-terpinene (12.4-14.0 per cent), thymol (29.7-35.1 per cent) and carvacrol (12.4-20.9 per cent) as the major constituents while oil of *O. vulgare* L. from Dharchula and Munsiyari (chemotype II) showed the presence of thymol (30.2-55.1 per cent), caryophyllene oxide (7.5-7.6 per cent) and aliphatic hydrocarbons (12.9-30.4 per cent). The oil from Ramgarh (chemotype III) represents linalool (10.9 per cent), thymol (5.1 per cent), carvacrol (7.5 per cent), β-caryophyllene (10.4 per cent) and germacrene D (5.7 per cent) as major constituents. Kilbury, Mukteshwar, Mussoorie, Nainital and Rushi village (chemotype IV) showed linalool (5.1-9.7 per cent), bornyl acetate (6.9-16.8 per cent), β-caryophyllene (9.2-16.7 per cent), germacrene D (5.7-13.0 per cent), bicyclogermacrene (0.3-5.4 per cent) and elemol (0.8-5.3 per cent) as the major constituents. Three chemotypes of *O. vulgare* L. were observed by De Martino *et al.* (2009): carvacrol/thymol (21.89 per cent/18.21 per cent); thymol/α-terpineol (26.75 per cent/15.10 per cent) and linalyl acetate/linalool (15.90 per cent/12.50 per cent) type.

The chemotypes are grouped according to the presence or absence of chemical markers (Figure 13.1).

Chemotype I: p-cymene, γ-terpinene, thymol and carvacrol.

Chemotype II: thymol, caryophyllene oxide and aliphatic hydrocarbons.

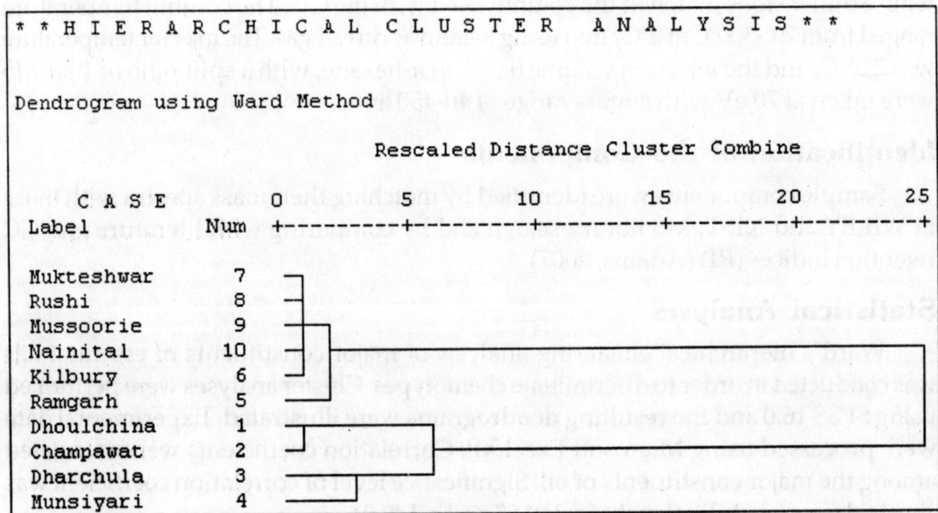

Figure 13.1: Agglomerative Hierarchical Clustering Analysis (Dendrogram) by SPSS 16.0 for the Chemical Abundances of 13 Essential Oil Components in the 10 Populations of *O. vulgare* L

Table 13.1: Chemotypes of *O. vulgare* Collected from Different Sites

Sl.No.	Compounds[a]		RI[b]	RI[c]	Chemotype I					Chemo-type II	Chemotype III		Chemotype II	
					Mukte-shwar (7)	Rushi Village (8)	Mus-soorie (9)	Nainital (10)	Kilbury (6)	Ram-garh (5)	Dhoul-china (1)	Cham-pawat (2)	Dhar-chula (3)	Munsi-yari (4)
1.	Santolinatriene	AH	908	906	0.3	0.4	–	–	–	–	–	–	–	–
2.	α-Thujene	MH	930	924	–	–	–	–	–	0.5	–	1.0	–	–
3.	α-Pinene	MH	939	932	0.7	0.9	–	–	0.2	0.6	t	0.4	–	–
4.	Camphene	MH	954	946	0.9	–	–	–	0.6	–	–	–	–	–
5.	Sabinene	MH	975	969	0.4	0.6	–	–	1.7	3.0	t	1.5	0.3	0.8
6.	β-Pinene	MH	979	974		–	1.9	–	2.7	–	t	–	–	0.2
7.	3-Octanone	AK	983	979	0.8	0.4	–	–	0.4	–	–	0.7	0.3	–
8.	β-Myrcene	MH	990	988	1.9	1.1	0.6	–	0.3	–	1.3	2.3	–	0.2
9.	3-Octanol	AA	991	988	0.2	0.4	–	–	3.7	–	–	0.3	–	–
10.	α-Phellendrene	MH	1002	1002		–	–	–	–	–	0.4	0.2	1.0	–
11.	α-Terpinene	MH	1017	1014	0.3	0.2	–	–	0.2	–	2.0	2.2	–	0.2
12.	p-Cymene	MH	1024	1020	0.3	0.5	–	–	0.9	–	**6.7**	**9.8**	–	t
13.	Limonene	MH	1029	1024	4.3	3.9	0.5	1.1	0.5	0.7	0.3	–	–	0.2
14.	1,8-Cineole	OM	1031	1026		–	–	4.5	3.4	2.6	1.1	0.2	0.9	–
15.	(Z)- β-Ocimene	MH	1037	1032		–	1.3	–	4.9	1.5	–	1.0	–	–
16.	(E)- β-Ocimene	MH	1050	1044	0.6	0.6	–	–	–	–	–	–	–	T
17.	γ-Terpinene	MH	1059	1054	3.0	3.6	1.4	4.0	0.6	1.5	**12.4**	**14.0**	0.3	–
18.	(Z)-Sabinene hydrate	MH	1070	1065	4.0	3.2	–	–	2.1	–	–	0. 8	0.8	0.5
19.	Linalool	OM	1096	1095	**8.5**	**6.7**	**5.1**	**7.7**	**9. 7**	**10.9**	4.0	1.3	t	0.2

Contd...

Table 13.1–Contd...

Sl.No.	Compounds[a]		RI[b]	RI[c]	Chemotype I					Chemotype II	Chemotype III		Chemotype II	
					Muktešhwar (7)	Rushi Village (8)	Mussoorie (9)	Nainital (10)	Kilbury (6)	Ramgarh (5)	Dhoulchina (1)	Champawat (2)	Dharchula (3)	Munsiyari (4)
20.	3-Octanolacetate		1123	1120	0.9	0.3	–	–	–	–	–	–	–	–
21.	Camphor	OM	1146	1141	–	–	–	–	–	–	–	–	1.4	0.4
22.	n.i.		–	–	0.3	0.6	–	–	–	–	–	0.3	–	–
23.	Borneol	OM	1169	1165	2.4	2.9	1.5	5.4	1.2	–	–	0.7	0.4	0.4
24.	Terpinen-4-ol	OM	1177	1174	0.8	0.4	–	–	1.8	–	0.6	–	–	0.6
25.	Methyl chavicol	OM	1196	1195	–	–	–	–	–	–	–	–	–	0.3
26.	α-Terpineol	OM	1188	1186	–	–	0.6	4.3	1.4	3.1	0.2	0.3	0.4	–
27.	Thymolmethyl ether	OM	1236	1232	–	–	–	–	–	–	0.2	3.1	0.3	0.9
28.	Carvacrol methyl ether	OM	1244	1241	–	–	–	–	–	–	4.5	–	t	0.4
29.	Cumin aldehyde	OM	1245	–	–	–	–	–	–	–	–	–	–	0.4
30.	Thymoquinone	OM	1252	1248	–	–	–	–	–	–	–	–	5.6	2.8
31.	Bornyl acetate	OM	1288	1287	**18.6**	**16.8**	**6.9**	**9.4**	**12. 6**	–	–	–	–	–
32.	Thymol	OM	1290	1289	–	–	–	–	–	**5.1**	**29.7**	**35.1**	**55.1**	**30.2**
33.	Carvacrol	OM	1299	1298	–	–	–	–	–	**7.5**	**20.9**	**12.4**	1.9	1.0
34.	Δ-Elemene	SH	1338	1335	–	–	–	–	–	–	0.3		–	–
35.	*trans*-Carvylacetate	OM	1342	1339	–		9.2	–		–				
36.	Thymylacetate	OM	1352	1349	–	–	–	–	–	–	t	–	–	–
37.	α-Copaene	SH	1376	1374	1.7	1.3	–	–	1.1	–	–	–	–	1.6
38.	β-Bourbonene	SH	1388	1387	–	–	1.8	2.6	1.5	3.1	–	–	–	–

Contd...

Table 13.1–Contd...

Sl.No.	Compounds[a]		RI[b]	RI[c]	Chemotype I					Chemo-type II	Chemotype III		Chemotype II	
					Mukte-shwar (7)	Rushi Village (8)	Mus-soorie (9)	Nainital (10)	Kilbury (6)	Ram-garh (5)	Dhoul-china (1)	Cham-pawat (2)	Dhar-chula (3)	Munsi-yari (4)
39.	β-Elemene	SH	1389	1388	–		2.2	1.9		3.2				
40.	(Z)-α-Bergamotene	SH	1412	1411	–	1.0	–	–	–	–	–	3.0	–	–
41.	(E)-Caryophyllene	SH	1419	1417	**14.3**	**10.9**	**9.2**	**16.7**	**13.8**	**10.4**	2.3	1.2	–	0.2
42.	Aromadendrene	SH	1441	1439	–		0.6	–		–				–
43.	α-Humulene	SH	1454	1452	0.6	0.9	–	2.8	2.0	1.7	t	–	–	–
44.	Germacrene D	SH	1485	1484	**13.0**	**11.3**	**18.0**	**11.7**	**6.3**	**5.7**	1.9	–	–	0.7
45.	Δ-Selinene	SH	1492	1492	0.4	1.3	–	–	–	–	–	–	–	–
46.	*epi*-Cubebol	OS	1494	1493	–	–	–	–	–	–	0.4	–	–	–
47.	Bicyclogermacree	SH	1500	1500	0.6	0.8	2.3	**5.4**	0.3	3.4				
48.	(E),(E)-α-Farnesene	SH	1505	1505	–	1.2	1.1	0.4		2.3	–	–	–	0.5
49.	β-Bisabolene	SH	1505	1505	5.3	4.7	2.1	3.9	3.2	–	–	–	–	–
50.	Δ-Cadinene	SH	1523	1522	–	–	–	–	3.1	–	2.8	–	–	–
51.	Elemol	OS	1549	1548	3.9	4.1	**5.3**	0.8	1.0	–	–	–	–	–
52.	*cis*-Muurol-5-en-4-2-ol	OS	1561	1559	–	–	–	–	2.1	–	–	–	–	–
53.	Germacrene D-4-ol	OS	1575	1574	–	–	–	–	–	–	t	2.5	–	–
54.	Spathulenol	OS	1578	1577	1.6	2.9	–	–	4.5	–	–	0.5	–	1.2
55.	Caryophyllene oxide	OS	1583	1582	1.4	2.6	**3.7**	**1.8**	0.9	**1.0**	0.4	–	**7.6**	**7.5**
56.	Globeulol	OS	1590	1590	–		1.0	–		–				
57.	10-*epi*-γ-Eudesmol	OS	1623	1622	–	–	1.1	–	–	–	0.2	–	–	–

Contd...

Table 13.1–Contd...

Sl.No.	Compounds[a]		RI[b]	RI[c]	Chemotype I					Chemotype II	Chemotype III		Chemotype II	
					Mukteshwar (7)	Rushi Village (8)	Mussoorie (9)	Nainital (10)	Kilbury (6)	Ramgarh (5)	Dhoulchina (1)	Champawat (2)	Dharchula (3)	Munsiyari (4)
58.	*epi*-α-Cadinol	OS	1640	1638	0.7	1.3	–	–	–	–	–	–	–	–
59.	α-Muurolol	OS	1646	1644	–		3.8	1.2		4.5				
60.	Cubenol	OS	1646	1645	–		0.7	1.7		–				
61.	β-Eudesmol	OS	1649	1650	–		0.9	–		–				
62.	α-Cadinol	OS	1654	1652	3.1	2.7	**8.8**	3.5	1.1	**9.3**	–	–	–	–
63.	α-Bisabolol	OS	1685	1685	1.3	1.2	–	–	1.4	1.4	0.4	–	–	–
64.	n-Octadecane	AH	1800	1800	–		0.6	0.3		0.8				
65.	n-Nonadecane	AH	1900	1900	–		0.4	–		–				
66.	Tricosane	AH	2300	2300	–	–	–	–	–	–	–	–	–	4.7
67.	Tetracosane	AH	2400	2400	–	–	–	–	–	–	–	–	3.8	**6.6**
68.	Hexacosane	AH	2600	2600	–	–	–	–	–	–	–	–	2.7	**8.0**
69.	Heptacosane	AH	2700	2700	–	–	–	–	–	–	–	–	2.8	**7.8**
70.	Octacosane	AH	2800	2800	–	–	–	–	–	–	–	–	3.5	**8.0**
	Percent of oil identified (per cent)				**97.1**	**91.7**	**92.6**	**91.1**	**91.2**	**83.8**	**91.2**	**93.0**	**94.8**	**91.1**

a: Mode of identification: Retention Index, coinjection with standards/Peak enrichment with known oil constituents.

b: Retention indices determined on the Equity-5 column using an *n*-alkane homologous series (C9–C24).

c: Retention indices from the literature (Adams, 2007).

Bold type indicates major components, per cent).

n.i.: Not identified; AK: Aliphatic ketone; AA: Aliphatic alcohol; MH: Monoterpene hydrocarbon; OM: Oxygenated monoterpene; SH: Sesquiterpene hydrocarbon; OS: Oxygenated sesquiterpene.

Within the brackets, numbers denotes the accession number in dendrogram in Figure 13.1.

Chemotype III: linalool, thymol, carvacrol, β-caryophyllene and germacrene D.

Chemotype IV: linalool, bornyl acetate, β-caryophyllene, germacrene D, bicyclogermacrene and elemol.

On the basis of biosynthetic pathway these four chemotypes were further classified into two subtypes. These two subtypes show distinct biogenetic pathways; Subtype-I is phenylpropanoid type belonging to shikimic acid pathway whereas Subtype- II is monoterpenoid type formed by the mevalonic acid pathway.

Subtype-I

Chemotype I: p-cymene, γ-terpinene, thymol and carvacrol.

Chemotype II: thymol, caryophyllene oxide and aliphatic hydrocarbons.

Chemotype III: linalool, thymol, carvacrol, β-caryophyllene and germacrene D.

Subtype-II

Chemotype IV: linalool, bornyl acetate, β-caryophyllene, germacrene D, bicyclogermacrene and elemol.

The chemical composition of essential oils isolated from aerial parts of *O. vulgare* from Northwestern Himalaya was investigated by Bisht *et al.* (2009). They have reported the presence of linalool (11.0-14.7 per cent), bornyl acetate (7.0-9.3 per cent) along with borneol (3.4-5.9 per cent), terpenen-4-ol (0.6-1.2 per cent) and α-terpeneol (2.4-8.4 per cent) in Nainital and Bhawali regions. While our reports were entirely different from the previous oil reports. In chemotype I, p-cymene, g-terpinene, thymol and carvacrol were the major constituents. Chemotype II was rich in thymol, caryophyllene oxide and aliphatic hydrocarbons. Linalool, thymol, carvacrol, β-caryophyllene and germacrene D were found the major constituents in chemotype III. Chemotype IV from Nainital was different from the earlier report as linalool (7.7 per cent), borneol (5.4 per cent), bornyl acetate (9.4 per cent), β-caryophyllene (16.7 per cent), germacrene D (11.7 per cent) and bicyclogermacrene (5.4 per cent) were the major constituents while terpinen-4-ol was totally absent. Therefore, chemotype I, II, III and IV show different chemical profile as compared to the earlier reports.

Correlation among Major Constituents

The statistically significant correlations within the major constituents of *O. vulgare* are given in Table 13.2. As shown in this table there is considerable number of significant correlations. p-Cymene is positively correlated with γ-terpinene and carvacrol ($r=0.950$, $Pd \leq 0.01$; $r=0.812$, $Pd \leq 0.01$, respectively). γ-Terpinene was found to be positively correlated with carvacrol ($r=0.826$, $P=<0.01$). Poulose and Croteau (1978) have reported that γ-terpinene and p-cymene are the biosynthetic precursors (via enzymatic hydroxylation) of the two isomeric phenols, carvacrol and thymol, in *Thymus vulgaris* essential oil. The variation of the concentration of thymol and carvacrol depends upon monoterpene hydrocarbons. In addition, some monoterpene hydrocarbons as a total (β-pinene, myrcene, camphene, α-terpinene, γ-terpinene and p-cymene) were found to be responsible for the variation of the concentration of linalool (Kanias *et al.*, 1998). In our study, p-cymene and γ-terpinene provide evidence that these monoterpenic hydrocarbons can be the biosynthetic precursor of carvacrol.

Table 13.2: Simple Correlation Matrix (r) among Major Constituents

Sl.No.	1 p-Cymene	2 γ-Terpinene	3 Linalool	4 Bornyl Acetate	5 Thymol	6 Carvacrol	7 (E)-Caryophyllene	8 Germacrene D	9 Bicyclegermacrene	10 β-Bisabolene	11 Elemol	12 Caryophyllene Oxide	13 Aliphatic Hydrocarbons
1.	1.00	0.950**	–0.366	–0.393	0.414	0.812**	–0.476	–0.491	–0.394	–0.444	–0.362	–0.450	–0.243
2.		1.00	–0.246	–0.269	0.275	0.826**	–0.317	–0.322	–0.210	–0.291	–0.255	–0.546	–0.383
3.			1.00	0.570	–0.851**	–0.217	0.883**	0.566	0.528	0.541	0.294	–0.605	–0.641*
4.				1.00	–0.729*	–0.580	0.775**	0.707*	0.110	0.949**	0.717*	–0.269	–0.410
5.					1.00	0.423	–0.908**	–0.829**	–0.545	–0.786**	–0.619	0.555	0.525
6.						1.00	–0.496	–0.530	–0.260	–0.626	–0.493	–0.390	–0.208
7.							1.00	0.763*	0.634	0.810**	0.477	–0.487	–0.583
8.								1.00	0.525	0.874**	0.879**	–0.362	–0.481
9.									1.00	0.299	0.097	–0.259	–0.336
10.										1.00	0.825**	–0.293	–0.443
11.											1.00	–0.217	–0.349
12.												1.00	0.877**
13.													1.00

*: Correlation is significant at the 0.05 level; **: Correlation is significant at the 0.01 level.

Linalool is positively correlated with β-caryophyllene (r=0.883, Pd≤0.01) while negatively correlated with thymol (r= -0.853, Pd≤0.01) and aliphatic hydrocarbons (r= -0.644, Pd≤0.05). Bornyl acetate in the plant essential oil was positively correlated with β-caryophyllene (r=0.775, Pd≤0.01), germacrene D (r=0.707, Pd≤0.05), β-bisabolene (r=0.949, Pd≤0.01), elemol (r=0.717, Pd≤0.05), and negatively correlated with thymol (r= -0.729, Pd≤0.05). Thymol was negatively correlated with β-caryophyllene, (r= -0.908, Pd≤0.01), germacrene D (r= -0.828, Pd≤0.01) and β-bisabolene (r= -0.783, Pd≤0.01). β-Caryophyllene was positively correlated with germacrene D (r=0.763, Pd≤0.05) and β-bisabolene (r=0.810, Pd≤0.01). Germacrene D was showing positive correlation with β-bisabolene (r=0.874, Pd≤0.01) and elemol (r=0.879, Pd≤0.01). β-bisabolene was positively correlated with elemol (r=0.825, Pd"0.01) while caryophyllene oxide showed positive correlation with aliphatic hydrocarbons (r=0.877, Pd≤0.01). Pluhár, *et al.* (2007) observed that in *Thymus pannonicus* All. and *Thymus praecox* Opiz, thymol concentration was positively correlated with nitrogen content of the soil.

Conclusion

The cluster analysis confirmed the presence of four chemotypes and on the basis of biosynthetic pathway these four chemotypes were further classified into two groups; phenylpropanoid group belonging to shikimic acid pathway whereas monoterpenoid group formed by the mevalonic acid pathway. Our result showed that p-cymene and γ-Terpinene were positively correlated with carvacrol. This is due to p-cymene and γ-Terpinene act as biosynthetic precursors (via enzymatic hydroxylation) of the two isomeric phenols, carvacrol and thymol.

Acknowledgement

The author is grateful to Director, USERC, Dehradun for financial support and Head, Chemistry Department, Kumaun University, Nainital for providing the necessary laboratory facilities and encouragement.

References

Adams, R.P. (2007). Identification of Essential Oil Components by Gas Chromatography/Mass Spectrometry (4th Ed.), Allured Pub. Corp., Carol Stream. IL.

Polunin, A., Stainton, A., 1984. Flowers of the Himalaya. Oxford University Press, New Delhi

Ietswaart, J. H., 1980. A Taxonomic Revision of the Genus *Origanum* (Labiatae). Leiden Univ. Press, The Hague/Boston/London, London.

Bodrug, M. V., 1982. Status and prospects of study and use of aromatic plants in the Moldavian-SSR, USSR. Rastit. Resur. 18, 558-560.

Chalchat, J. C., Pasquier, B., 1998. Morphological and chemical studies of *Origanum* clones: *Origanum vulgare* L. ssp. *vulgare.* J. Essent. Oil Res. 10, 119-125.

Russo, M., Galletti, G. C., Bocchini, P., Carnacini, A., 1998. Essential oil chemical composition of wild populations of Italian oregano spice (*Origanum vulgare* ssp.

hirtum (Link) Ietswaart): a preliminary evaluation of their use in chemotaxonomy by cluster analysis. 1. Inflorescences. J. Agric. Food Chem. 46, 3741-3746.

D'antuono, L. Filippo, Galletti, Guido C., Bocchini, Paola., 2000. Variability of essential oil content and composition of *Origanum vulgare* L. populations from a north Mediterranean area (Liguria Region, Northern Italy). Annals of Bot. 86, 471-478.

Mockute, D., Bernotiene, G., Judzentiene, A., 2001. The essential oil of *Origanum vulgare* L. ssp. *vulgare* growing wild in Vilnius district (Lithuania). Phytochemistry 57, 65.

Baser, K. H. C., Ozek, T., Kurkcuoglu, M., Tumen, G., 1994. The essential oil of *Origanum vulgare* subsp. *hirtum* of Turkish origin. J. Essent. Oil Res. 6, 31-36.

Veres, K., Varga, E., Dobos, A., Hajdú, Z. S., Máthé, I., Németh, E., Szabó, K., 2003. Investigation of the composition and stability of the essential oils of *Origanum vulgare* ssp. *vulgare* L. and *O. vulgare* ssp. *hirtum* (Link) Ietswaart. Chromatographia 57, 95–98.

De Martino, Laura, De Feo, Vincenzo, Formisano, Carmen, Mignola, Enrico, Senatore, Felice, 2009. Chemical composition and antimicrobial activity of the essential oils from three chemotypes of *Origanum vulgare* L. ssp. *hirtum* (Link) Ietswaart growing wild in Campania (Southern Italy). Molecules 14, 2735-2746.

Kaul, V. K., Singh, B., Sood, R. P., 1996. Essential oil of *Origanum vulgare* L. from North India. J. Essent. Oil Res. 8, 101–103.

Pande, C., Mathela, C. S., 2000. Essential Oil Composition of *Origanum vulgare* L. from the Kumaon Himalayas. J. Essent. Oil Res. 12, 441-442.

Bisht. D., Chanotiya, C. S., Rana, M., Semwal, M., 2009. Variability in essential oil and bioactive chiral monoterpenoid compositions of Indian oregano (*Origanum vulgare* L.) populations from northwestern Himalaya and their chemotaxonomy. Ind. Crop Prod. 30, 422-426.

Schearer, W.R., 1984. Components of oil of tansy (*Tanacetum vulgare*) that repel Colorado potato beetles (*Leptinotarsa decemlineata*). J. Natur. Product. 47, 964-969.

2013, Impact of Global Climate Change on Earth Ecosystems *Pages* ***115–119***
Editors: **D.R. Khanna, A.K. Chopra, Gagan Matta, Vikas Singh & Rakesh Bhutiani**
Published by: **BIOTECH BOOKS, NEW DELHI**

Chapter 14

Terpenoid Composition of Leaves, Seeds and Roots of *Cyclospermum leptophyllum*: A Comparative Study

Charu Singh, Chitra Pande, Geeta Tewari and Shalini Singh

Department of Chemistry, Kumaun University, Nainital – 263 002 Uttarakhand

The essential oil of Cyclospermum leptophyllum (Pers.) Sprague ex Britton and P. Wilson syn. *Apium leptophyllum* (Pers.) F. Muell. Ex Benth., family Apiaceae (Umbellifereae) is known by the common names marsh parsley and fir-leafed celery.

The fresh plant material of *C. leptophyllum* was collected from Nainital (Uttarakhand). A comparative study of essential oils extracted by steam distillation from leaves, seeds and roots of *C. leptophyllum* was done by GC and GC/MS. The major components were isolated by column chromatography.

The essential oils showed the presence of 26-32 compounds, of which 13-18 compounds were identified representing 96.2 per cent - 97.8 per cent of the total oil. The essential oil of *C. leptophyllum* mainly consists of oxygenated monoterpenes (68.7 per cent, 86.8 per cent and 92.5 per cent in leaves, seeds and roots respectively).

Our study showed the presence of thymohydroquinone dimethyl ether (50.7 per cent, 60.4 per cent and 71.7 per cent in leaves, seeds and roots respectively) followed by thymol methyl ether (11.2 per cent, 11.5 per cent and 13.7 per cent in leaves, roots and seeds respectively), γ-terpinene (10.4 per cent leaves), p-cymene (3.5 per cent and 9.5 per cent in seeds and leaves respectively) and carvacrol methyl ether (5.7 per cent, 5.9 per cent and 7.9 per cent in roots, leaves and seeds respectively) as predominant components. To the best of our knowledge this is the first report on the essential oil composition of *C. leptophyllum* leaves, seeds and roots from India.

Keywords: *Cyclospermum leptophyllum, Essential oil composition, Thymohydroquinone dimethyl ether, Thymol methyl ether.*

Introduction

Cyclospermum leptophyllum (Pers.) Sprague ex Britton and P. Wilson syn. *Apium leptophyllum* (Pers.) F. Muell. ex Benth. (*A. tenuifolium* and *C. ammi.*) belongs to the family Apiaceae (Umbellifereae) and is known by the common names marsh parsley and fir-leafed celery. In Hindi, it is known as "Ajmoda". The plant is native to America but found worldwide from warm temperate to tropical latitudes, and is considered as a noxious weed in many areas. It is a tap rooted branching herb reaching just over half a meter, and has thread like green leaves (Pullaiah, 2006). In Ethiopian traditional medicine, the leaves of A. leptophyllum (Pers.) Benth. are used for the treatment of a disease condition locally known as "Mitch" characterized mainly by inflammation, sweat and loss of appetite. The essential oil from the leaves of *C. leptophyllum* was found to be rich in thymohydroquinone dimethyl ether (*Asamenew, Tadesse, Asres, Mazumder, and Bucar, 2008;* Park, and Sutherland, 1969). *Ayapana triplinervis* and *Sphaeranthus suaveolens* DC are also a good source of thymohydroquinone dimethyl ether (Gauvin-Bialecki and Marodon, 2008; De Pooter, Buyck, Schamp, Harraz, and EI-Shami, 1991). The essential oil from the flowers has a high percentage of thymohydroquinone dimethyl ether and showed strong antifungal, antibacterial, antiparasitic and anthelminitic activity (Sharma, 1979; Garg, 1993). Essential oil of *C. leptophyllum* is medicinally used as a mild balm, in stomachic and diarrhoea (Everitt, Lonard, and Little, 2007).

The aim of this work was to investigate the essential oil composition of *C. leptophyllum* leaves, roots and seeds from Uttarakhand.

Material and Methods

Plant Collection and Identification

The fresh leaves, roots and seeds of *C. leptophyllum* were collected from Nainital (Uttarakhand). The plant was identified by Prof. Y. P. S. Pangtey, Botany Department, Kumaun University, Nainital, and is deposited in the herbarium of Botanical Survey of India, Dehradun (Voucher no. BSD-112180).

Extraction of Essential Oil

The fresh leaves, roots and seeds of *C. leptophyllum* were subjected to steam distillation using a copper electric still, fitted with spiral glass condensers. The distillates were saturated with NaCl and extracted with *n*-hexane and dichloromethane. The organic phase was dried over anhydrous sodium sulfate and the solvents were distilled off in a rotary vacuum evaporator at 30°C and the percentage oil content was calculated on the basis of fresh weight of plant materials. The oil yield was 0.3 per cent (v/w) in leaves, 0.2 per cent (v/w) roots and 0.3 per cent (v/w) seeds.

Analysis of the Essential Oil

The oil was analyzed by using a Perkin Elmer Autosystem XL GC. The column temperature was programmed from 70°C–290°C at 3°C/min using hydrogen as the carrier gas at 10 psi column head pressure, The injector temperature was 280°C and detector (FID) temperature was 290°C. The GC/MS used was Autosystem XL GC fitted with Equity-5 (60m x 0.32mm i.d., film thickness 0.25µm) fused silica capillary column coupled with Perkin Elmer Turbomass fitted with Equity-5 capillary column whose dimensions matched the column used with the GC. The column temperature ranged from 70°C–300°C at 3°C/min using helium as carrier gas. The injector temperature was 220°C, and the injection volume 0.05 L in *n*-hexane, with a split ratio of 1:50. MS were taken at 70 eV with a mass range of 40-450 amu.

Identification and Isolation of the Ccomponents

Identification of the constituents was done on the basis of Retention Index, MS Library search (NIST and WILEY) and by comparing with the MS literature data (Adams, 2007). The relative amounts of individual components were calculated based on GC peak area (FID response) without using correction factor. Isolation of compounds was done by using column chromatography and identification was done by using Co-injection with authentic sample in GC.

Results and Discussion

The leaves, roots and seeds of *C. leptophyllum* on steam distillation gave light yellow colored essential oils (Yield: 0.3 per cent (v/w) in leaves, 0.2 per cent (v/w) roots and 0.3 per cent (v/w) seeds). The compounds were identified on the basis of correlation of their mass spectra and comparison with those given in the literature (Adams, 2007) and major constituents were identified by Co-injection with authentic samples. GC and GC/MS analysis of essential oil of *C. leptophyllum* collected from Nainital (Uttarakhand) showed the presence of 32 compounds, of which 20 compounds were identified, which represent 98.07 per cent of the total oil in leaves, 95.30 per cent roots and 96.20 per cent seeds. Earlier reports showed the presence of thymohydroquinone dimethyl ether, isothymol methyl ether, thymol methyl ether as major constituents (Khare, 2004; Park, and Sutherland, 1969). Our study shows the presence of thymohydroquinone dimethyl ether (50.7 per cent), thymol methyl ether (11.2 per cent), γ-terpinene (10.4 per cent), p-cymene (9.5 per cent) and carvacrol methyl ether (5.9 per cent) as major constituents. Thymohydroquinone dimethyl ether (46.2-50.7 per cent), isothymol methyl ether (10.6-11.5 per cent), thymol methyl ether (7.3 per cent), p-cymene (5.7-10.7 per cent) and γ-terpinene (3.1-6.2 per cent) were identified as the major constituents in the leaves essential oil of *Apium leptophyllum* (Shin'ichi, and Kayo, 2002). Our study has indicated the presence of significant quantity of carvacrol methyl ether (7.5 per cent - 5.9 per cent).

Table 14.1: Essential Oil Composition of Aerial Parts of *Cyclospermum leptophyllum*

*Compounds**	*RI*	*Nainital (Leaf Oil per cent)*	*Seeds Oil per cent*	*Roots Oil per cent*
α-thujene	930	0.1	–	0.1
α-pinene	939	0.2	–	t
sabinene	975	0.6	0.2	t
β-pinene	979	1.8	1.7	–
β-myrcene	990	0.9	0.1	–
mesitylene	995	–	–	0.3
α-phellandrene	1002	–	–	–
p-cymene	1024	9.5	3.5	0.1
β-phellandrene	1029	0.2	–	–
limonene	1029	0.6	–	–
cis-ocimene	1037	t	–	–
trans-ocimene	1050	2.5	3.9	0.1
γ-terpinene	1059	10.4	–	–
α-terpineole	1188	–	–	3.0
thymol methyl ether	1236	11.2	13.7	11.5
carvacrol methyl ether	1244	5. 9	7.9	5.7
cumin aldehyde	1245	0.9	–	–
p-ethylacetophenone	1281	–	4.2	0.6
thymol	1290	–	0.1	–
γ-terpinen-7-al	1291	–	0.3	–
thymohydroquinone dimethyl ether	1418	50.7	60.4	71.7
cis-methyl isoeugenol	1451	–	0.2	–
α-humulene	1454	0 9	–	–
germacrene D	1485	1.2	–	t
caryophyllene oxide	1583	0 2	–	0.7
n-octadecane	1800	–	–	1 6
Per cent oil identified		98 07	96.2	95 3

* Mode of identification: Retention Index, coinjection with standards with Peak enrichment of the oil constituents, MS (GC/MS), t = trace (< 0.1 per cent).

The essential oil composition of seeds oil showed thymohydroquinone dimethyl ether (60.4 per cent), thymol methyl ether (13.7 per cent), and carvacrol methyl ether (7.9 per cent) as major constituents while p-cymene, trans-ocimene, p-ethylacetophenone as minor constituents. The essential oil of roots oil was found to be rich in thymohydroquinone dimethyl ether (71.7 per cent), thymol methyl ether (11.5 per cent), and carvacrol methyl ether (5.7 per cent).

Conclusion

Essential oil of all parts of C. *leptophyllum* were rich in thymohydroquinone dimethyl ether, thymol methyl ether and carvacrol methyl ether. Comparative result of the volatile constituents of leaves, roots and seeds showed that there are some quantitative and qualitative differences in their essential oils. From the above results, it is concluded that the essential oil of C. *leptophyllum* can prove to be a useful source of thymol ethers. To the best of our knowledge, this is the first report on the essential oil composition of C. *leptophyllum* collected from India.

Acknowledgments

The authors are grateful to Director, USERC, Dehradun for financial support and Head, Chemistry Department, Kumaun University, Nainital for providing the necessary laboratory facilities and encouragement.

References

Adams, R. P. (2007) *Identification of essential oil components by Gas Chromatography/ Mass Spectroscopy* (4th Ed). Carol Stream. IL: Alllured Publishing Corporation.

Asamenew, G., Tadesse, S., Asres, K., Mazumder, A. and Bucar, F. (2008) Ethiopian Pharmaceutical Journal *26, 95-102.*

De Pooter, H. L., De Buyck, L. F., Schamp, N. M., Harraz, F. M. and El-Shami, I. M. (1991) *Flavour and Fragrance Journal* 6, 157-159.

Everitt, J. H. Lonard, R. L. and Little, C. R. (2007) *Weeds in South Texas and Northern Mexico*. Lubbock: Texas Tech University Press.

Garg, S.C. (1993) *Indian perfumer* 37, 318-323.

Gauvin-Bialecki, A. and Marodon, C. (2008) *Biochemical Systematics and Ecology* 36, 853-858.

Khare, C. P. (2007) *Encyclopedia of Indian Medicinal Plants: Rational Western therapy, ayurvedic and other traditional usage, Botany*. Berlin, Heidelberg: Springer-Verlag.

Park R. J. and Sutherland, M. D. (1969) *Australian Journal of Chemistry* 22, 495-496.

Pullaiah, T. (2006). *Encyclopedia of world medicinal plants* (Vol. 1). New Delhi, India: Regency Publications.

Sharma, S.K. (1979) *Indian Drugs of Pharmaceutical Industries* 1, 3-6.

Shin'ichi, F. and Kayo, M. (2002) *Koryo, Terupen oyobi Seiyu Kagaku ni kansuru Toronkai Koen Yoshishu* 46, 19-21.

2013, Impact of Global Climate Change on Earth Ecosystems *Pages* ***121–124***
Editors: **D.R. Khanna, A.K. Chopra, Gagan Matta, Vikas Singh & Rakesh Bhutiani**
Published by: **BIOTECH BOOKS, NEW DELHI**

Chapter 15

Transmission Electron Microscope Study of Erythrocytes in Patients of Fluorosis

A. Shashi and J. Sonam

Department of Zoology,
Punjabi University, Patiala – 147 002, Punjab

The aim of the present study was to ascertain the effect of fluoride on ultrastructure of red blood cells in patients aged 20-50 exposed to high fluoride in drinking water (8.05 to 10.25 mg/L, mean 9.15±1.55 mg/L) in two high fluoride areas of Bathinda district, Punjab. Blood samples of these patients were taken into EDTA coated vacutainers for transmission electron microscopy study of erythrocytes. Transmission electron microscope examination of red blood cells revealed hemolysis in erythrocytes. The red blood cells were swollen, membrane broken, and haemoglobin leaked out. Erythrocytes with vacuoles formation were observed. Exposure to fluoride caused formation of blister erythrocyte that has discrete areas of the cell membrane to become ill- defined and less dense. Some erythrocytes in fluorotic patients were predisposed to plasma membrane defects. These cells contained surface membrane remnants on erythrocyte. Red blood cells were transformed into degmacytes with one or two bites on membrane surface, dacrocyte, schizocyte and drepanocyte. Erythrocytes showed agglutination that represents clumping of an individual's red blood cells and some erythrocytes showed fusion through a bridge formation. Our study has confirmed that fluoride is one of the chemical substances responsible for producing different types of ultrastructural alterations in red blood cells in patients afflicted with fluorosis.

Keywords: Agglutination, Erythrocytes, Fluorosis, Hemolysis, Ultrastructural alterations.

Introduction

Fluoride is a chemical element that is very abundant on the earth's surface and that is recognized as an element that can produce acute and chronic changes in humans when they are exposed to it. Fluorosis is caused by acute or chronic ingestion of fluorides and is clinically characterized in humans by dental alterations, as well as alterations in the nervous and musculoskeletal systems (Foulkes, 2002). Fluoride can cross cell membranes and enter bone, soft tissues and blood (Carlson *et al.*, 1960; Jacyszyn and marut, 1986; Shivarajashankara *et al.*, 2001).

The present study has been undertaken to elucidate the ultrastructural alterations and modifications of red blood cells due to fluoride by using transmission electron microscopy in human subjects afflicted with dental and skeletal fluorosis residing in two endemic fluorotic areas (Daulatpura and Sivian) of Bathinda district of Punjab, India.

MATERIALS AND METHODS

Two villages (Daulatpura and Sivian), were selected for study in district Bathinda, Punjab, India. The level of fluoride in drinking water varied from 8.05 to 10.25 mg/L (mean 9.15 ± 1.55 mg/L) which was much higher than permissible fluoride levels (1.5 mg/L) as per WHO guidelines (2006). The study was approved by the Institutional ethics committee, Punjabi university, Patiala. For transmission electron microscopy, blood samples of fluorotic patients anticoagulated with EDTA were used and centrifuged at 1000 rpm for 10 minutes. The washed pellets were divided into fragments approximately 1 mm in diameter. These were dehydrated through graded concentrations of ethyl alcohol, embedded in Arladite, cut into 40 to 60 nm sections, and placed on parlodion- coated grids, stained with uranyl acetate and lead citrate and examined with a Philips CM-10 transmission electron microscope photomicrography.

Results and Discussion

In the present study, transmission electron microscope examination of red blood cells revealed hemolysis in erythrocytes, erythrocytes with vacuoles formation and blister formation. Some erythrocytes in fluorotic patients were predisposed to plasma membrane defects. These cells contained surface membrane remnants on erythrocytes. Red blood cells were transformed into degmacytes with one or two bites on membrane surface, dacrocytes, schizocytes and drepanocytes. Strunecka *et al.* (1991) observed shape changes after addition of 1mM sodium fluoride and 10 μM aluminium chloride in human red blood cells. Erythrocytes from patients with various disorders of lipoprotein metabolism have been found to have abnormal morphology. Bouaziz *et al.* (2006) reported that fluoride causes disruptive effect on erythropoiesis, enhanced production of superoxide radicals and lipid peroxidation that lead to alterations in erythrocytes cell membrane function and structure. Blood from cases of fluorosis contained various shape changes and transformed into spherocytes and schizocytes. There are multiple causes of spherocyte formation. Spherocytes are also associated with immune-mediated hemolytic anemias. However, the presence of spherocytes on a blood smear is not pathognomonic for this immune-mediated disease. Other pathological disease states can affect erythrocytes and cause spherocyte formation through defective membrane assembly and toxic injury to erythrocytes. The degmocytes with bites result from the removal of denatured hemoglobin by macrophages in the spleen. Glucose-6-phosphate dehydrogenase deficiency, in which uncontrolled oxidative stress causes hemoglobin to denature, disorder that leads to the formation of bite cells. Hemolytic anemia associated with severe liver disease is another setting where bite cells are formed. Erythrocytes showed agglutination that represents clumping of an individual's red blood cells and some erythrocytes showed fusion through a bridge formation. Fluoride ions can inhibit or activate various functions in blood cells (Bober *et al.*, 2001). The rapidity of hemolysis suggests that the action of the fluoride toxin itself lead to red cell destruction. High fluoride concentration may disturb the anion channel of the erythrocytes membrane, which leads to hemolysis and swelling of cells (Guminska and Sterkowicz, 1976; Guminska and Skowron-Sula, 1985; Grabowska *et al.*, 1991). Verma *et al.* (2006) demonstrated that various concentration of sodium fluoride in the range of 50 to 500 μg/mL cause destabilization of red blood cell membrane leading to influx of water into the cells thereby causing hemolysis. Our study has confirmed that fluoride is one of the chemical substances responsible for producing different types ultrastructural alterations in blood cells in patients afflicted with fluorosis.

References

Bober, J., Chlubek, D., Kwiatkowska, E., Kedzierska, K., Stachowska, E., Wieczorek, P., Byra, E., Machoy, Z. and Herdzik, E. 2001. *Fluoride*, 34(3): 174-180.

Bouaziz, H., Fetoui, H., Ketata, S., Jammoussi, K., Ellouze, F. and Zeghal, N. 2006. *Fluoride*, 39(3): 211-219.

Carlson, C. H., Armstrong, W. D., Singer, L. 1960. *Proc. Soc. Exp. Biol. Med.*, 104: 235-239.

Foulkes, R. G. 2002. *Fluoride.,35, 213-227.*

Grabowska, M., Guminska, M. and Ignacak, J. 1991. *Folia Med.Cracov.*, **32**(1): 103-110.

Guminska, M. and Skowron-Sula, M. 1985. *Folia. Med. Cracov.*, **26**: 35-41.

Guminska, M. and Sterkowicz, J. 1976. *Acta Biochem. Pol.*, **23**(4): 285-291.

Jacyszyn, K. and Marut, A. 1986. *Fluoride*, **19**: 26-36.

Shivarajashankara, Y. M., Shivashankara, A. R., Bhat, P. G., Rao, S. H. 2000. *Fluoride*, 33(2):66-73.

Strunecka, A., Dessouki N. I., Palecek, I., Kmonickowa, E., Krpezsova, L. and Potter, B.V. 1991. *Receptor*, **1**(3): 141-154.

Susheela, A. K. and Jain, S. K. 1986. *Stud. Env. Sci.*, **27**: 231-239.

Susheela, A. K., Sharma Y. O., Mohan, J., Singh, M, Jagannath, B. and Jain, S. K. 1982. *Fluoride* **15**(4): 173-177.

Verma, R. J., Trivedi, M. H. and Chinoy, N. J. 2006. *Fluoride*, **39**(4): 261-265.

World Health Organization. 2006. IWA Publishing London, Geneva. pp 1-131.

2013, Impact of Global Climate Change on Earth Ecosystems *Pages 125–129*
Editors: **D.R. Khanna, A.K. Chopra, Gagan Matta, Vikas Singh & Rakesh Bhutiani**
Published by: **BIOTECH BOOKS, NEW DELHI**

Chapter 16

Haematological Indices in Fluoride Toxicity

A. Shashi and G. Meenakshi

Department of Zoology,
Punjabi University, Patiala – 147 002, Punjab

The effect of fluorosis on haematological parameters and its relation to pathology of erythrocytes was investigated in two highly fluoride areas of Bathinda district, Punjab, India, where the level of drinking water fluoride varied from 8.05- 10.25 mg/L. The serum was analyzed for serum fluoride and uric acid. The serum fluoride levels were significantly ($P<0.0001$) higher in fluorotic patients, while serum uric acid content was significantly declined. The analysis of haematological parameters in male and female patients of fluorosis demonstrated significant ($P<0.0001$) decline in hemoglobin, red blood cell counts and packed cell volume. Pearson's correlation coefficient analysis revealed significant positive correlation of haemoglobin with erythrocyte count ($r=1.00$, $p<0.05$) and packed cell volume ($r=0.96$, $p<0.05$) and negative correlation with mean corpuscular volume ($r=0.92$, $p<0.05$) and mean corpuscular heamoglobin concentration ($r=0.64$, $p<0.05$). These changes in fluorotic patients are attributed to rate of bioaccumulation of fluoride in blood which further leads to a marked variation in structural morphology. A significant ($P<0.0001$) decrease in hemoglobin concentration was observed, which was reflective of the decrease in hematocrit and red blood cell counts that leads to anemia production with prominent poikilocytosis and other morphologies such as red cell inclusions, howel-Jolly bodies, hemoglobin c crystals, cabot's rings, normoblast, selenoid bodies and marked fusions through double bridge formation and multiple fusions in red blood cells in patients of fluorosis. Anemia vary from mild to moderately severe. The presence of nucleated red blood cells in peripheral blood signals a regenerative response by marrow to anemia, indicating markedly accelerated erythropoiesis and severe bone marrow stress indicated by decreased values of erythrocyte counts and haemoglobin in patients of fuorosis. Our results indicate that fluoride play an important role in several pathologies, which could be related to its interaction with cell membranes and has adverse effect on hematological indices.

Keywords: *Mean corpuscular heamoglobin, Mean corpuscular haemoglobin concentration Packed cell volume, Poikilocytosis, Red cell inclusions and fusions.*

Introduction

Fluoride is a major environmental contaminant, the impact of which on various human and animal tissues and organs has been studied extensively (Whitford, 2000). Adverse hematological effects of fluoride have also been reported including damage to hematopoietic organs (Machalinski *et al.*, 2000; Eren *et al.*, 2005; Ersoy *et al.*, 2010; Mythili and Devi, 2010). Fluorosis is endemic in many places around the world such as India, China, and Africa (Wang and Huang, 1995; Karadeniz and Altintas, 2008). Punjab is the one the Northern state affected with hydro fluorosis (Shashi and Bhardwaj, 2011). Fluoride intoxication has a toxic effect on clonogenicity of human progenitor cord blood cells (Zejmo *et al.*, 1998) and bone marrow hematopoietic cells (Machalinski *et al.*, 1999).

Although there are numerous studies about effects of fluoride on hematological parameters in experimental animals, there are limited studies about effects of chronic fluorosis on hematological parameters in human beings with endemic fluorosis in clinical settings.

Materials and Methods

Study Design

The present study was conducted in two highly fluorotic areas Daulatpura and Sivian of district Bathinda, Punjab, India. The study group consisted of 140 patients affected with fluorosis (68 males, 72 females, mean age 42.12 ± 11.94 years) consuming water with fluoride levels 8.05 to 10.25 mg/L (mean 9.15 ± 1.55 mg/L). Endemic fluorosis was diagnosed according to the clinical diagnosis criteria, as described by Teotia *et al.* (1985).

Haematological Analysis

Blood samples of fluorotic patients were taken into EDTA coated vacutainers to study morphology and haematological indices. The total erythrocytic count, haemoglobin concentration, haematocrit, mean corpuscular volume, mean corpuscular hemoglobin and mean corpuscular hemoglobin concentration.

Biochemical Analysis

The serum was separated by centrifugation (3000 rpm) for 10 minutes at room temperature for detection of serum fluoride (Ion selective electrode method (Hardwood, 1969) and uric acid (URICASE-POD method (Barham and Trinder, 1972).

Light Microscopic Examination

The blood films were stained by May-Grünwald's Giemsa staining method for light microscopic examination. Differentiation was done in buffered distilled water. Differential morphological analysis of blood cells were done under oil immersion by research microscope (Motic). The microphotographs were taken with a digital camera (Motic-480) attached on inverted research binocular microscope.

Statistical Analysis

The results were given as Mean ± SD, hematological values between males and females were compared by t – test, using statistica (version 7.0) and level of significance set at P< 0.0001. Pearson's correlation coefficient test was used to measure the correlation between fluorosis and hematological parameters.

Results and Discussion

The fluorotic patients of both sexes showed highly significant (p<0.0001) abnormalities in hematological parameters, the levels of hemoglobin, red blood cell counts and hematocrit values in female fluorotic patients, were significantly decreased. The fluoride induced anemia observed in this study may result from inhibitition of globulin synthesis (Obrig *et al.*, 1971), depression of erythropoiesis or a decrease in the level of blood folic acid (Cetin *et al.*, 2004; Hoogstratten *et al.*, 1965). Choubisa (1996) reported a decrease in red blood cell counts and hemoglobin in an area of endemic fluorosis. The decreased hematocrit levels may be attributed to a decrease in size of erythrocytes due to stressful conditions (Teotia *et al.*, 1981). Mythili and Devi, (2010) reported that fluoride toxicity affects the hemoglobin levels, red cell distribution width, mean corpuscular volume, significantly; red blood cell count exhibits small variations in few patients. A significant decrease in hemoglobin concentration was observed which was reflective of the decrease in hematocrit and red blood cell counts leads to anemia production (howell jolly bodies, cabot's rings, nucleated red blood cells, and reticulocytes). Changes of these values in fluorotic patients have been attributed to rate of bioaccumulation of fluoride in blood which further leads to a marked variation in structural morphology.

The values of Mean corpuscular volume, Mean corpuscular hemoglobin and Mean corpuscular hemoglobin concentration showed non-significant alterations. The significant (P<0.001) decline in hemoglobin concentration was reflective of the decrease in hematocrit and red blood cell counts. Correlation matrix of the data showed highly significant (P<0.05) positive correlation between hemoglobin and erythrocyte count (r=1.00). Hematocrit value showed highly significant positive correlation with hemoglobin (0.96) and erythrocyte count (r=0.96). MCH showed positive correlation with MCV (r=0.79) and highly significant negative correlation with hemoglobin (r=-0.92), erythrocyte count (r=-0.91) and hematocrit (r=-0.87) while MCHC showed a positive correlation with MCH (r=0.77) and significant negative correlation with hemoglobin (r=-0.64), erythrocyte count (r=-0.63) and hematocrit (r=-0.76). Serum fluoride levels were significantly higher (P < 0.0001) in all age groups of fluorotic patients. One way ANOVA demonstrated a highly significant (F = 100.379, P < 0.001) increase in serum fluoride concentration in fluorotic patients of all fluoride exposed groups. Serum uric acid levels were significantly ($P < 0.001$) decreased. One way ANOVA demonstrated a highly significant ($F = 8.426$, $P<0.005$) increase in serum uric acid concentration in fluorotic patients of all fluoride exposed groups.

Microscopic images of peripheral blood smear from fluorotic patients revealed the presence of various types of deformed red blood cells. The term 'poikilocytosis' describes a marked variation in the shape of erythrocytes, In cases of fluorosis, various forms of poikilocytes such as comma-shaped, bell-shaped, bean-shaped, bulb-shaped, and pincers (mushroom-shaped) were observed. Howel-Jolly bodies contained small, round, dense, dark violet granules. One or two howell jolly body per cell was observed in blood smear preparation of fluorotic patients. In fluorotic cases, hemoglobin C Crystals were present as darkly stained rhomboidal shapes in the red cells. Cabot's rings were darkly black stained thread like filaments present in the shape of a ring in the red cells. These are thought to be formed from the microtubules of the mitotic spindle or remnants of the nuclear membrane. Immature erythrocytes, when stained with May- Grünwald giemsa stain, exibit a darkly stained reticulum. These dark staining clumps of reticulum were the identifying feature of the reticulum. The amount of reticulum in a reticulocyte varies from a large clump in the most immature cells called group 1 reticulocytes to a network of reticulum called group 2 reticulocytes were prominent in blood of fluorotic patients. Normoblasts were very immature forms of red blood cells seen when there is a severe demand for red blood cells to be released by the bone marrow, appeared in blood of fluorotic patients. The presence of nucleated red blood cells in peripheral blood signals a regenerative response by marrow to anemia, indicating markedly accelerated erythropoiesis and severe bone marrow stress. Elevated numbers of reticulocytes is associated with hypoxia, red blood cell destruction, glucose-6-phosphate dehydrogenase deficiency and haemolytic anemia. Selenoid bodies were faintly staining red blood cells having large, rounded bodies with pink crescent like margins. Fluoride is capable of causing anemia, which could directly be related to the production of anemic cells such as selenoid bodies, and red cells inclusions – cabot's rings and howel jolly bodies which have also been reported in other anemic conditions (Jain, 1993). A number of red cells fused with each other and fusion occurred between two cells with direct dissolution of walls of both cells either through formation of double bridge and multiple red cell fusions were also noted.

The present study showed a significant positive correlation between blood fluoride levels, erythrocyte indices and morphological changes. Hence it provides further evidences that the blood cells membrane is the primary site of action of fluoride intoxication. Structural alterations correlated well with the kinetics of the fluoride toxin.

References

Barham, D. and Trinder, P. 1972 -Analyst, 97: 142.

Cetin, N., Bilgili, A., Eraslan, G. and Koyu, A. 2004. *Eur. J. Health Sci.,* 13: 46-50.

Choubisa, S.L. 1996. *Indian J. Env. Health,* 38: 119-126.

Eren, E., Ozturk, M., Mumcu, E.F. and Canatan, D. 2005. *Toxicol. Indian Health,* 21: 255-258.

Ersoy, I.H., Alanoglu, *E.G.*, Koroglu, B.K., Varol, S., Akcay, S., Ugan, Y., Ersoy, S. and Tamer, M.N. 2010. *Biol. Trace Elem. Res.*, 133(2): 121-127

Harwood, J. E. 1969. *Water Res.*, 3: 273-280.

Hoogstratten, B., Leone, N.C., Shupe, J.L., Greenwood, D.A. and Lieberman, J. 1965. *J. Am. Vet. Med. Assoc.*, 192(1): 26-32.

Jain, N.C. 1993. *Essentials of veterinary hematology*. Williams and Willkins, US, pp 133-158.

Karadeniz, A. and Altintas, L. 2008. *Fluoride*, 41(1): 67-71.

Teotia, S.P.S. Teotia, M. and Singh, R.K. 1981. *Fluoride*, 14(2): 69-74.

Machalinski, B., Dziedziejko, V. and Stecewicz, I. 1999. Preliminary report. *Metabolism Fluoru.*, 9: 83-87.

Machalinski, B., Zejmo, M., Stecewicz, I., Machalinska, A., Machoy, Z. and Ratajczak, M.Z. 2000. *Fluoride*, 33 (4): 168-173.

Mythili, K. and Devi, K.S. 2010. *The Bioscan*. **5**(2): 255-258.

Obrig, T., Irvin, J., Culp, W.and Hardesty, B. 1971. *Eur. J. Biochem.*, 21(1): 31-41.

Shashi, A., and Bhardwaj, M. 2011. *Asian Journal of Water, Environment and Pollution*, 8(2):137-142.

Teotia, S.P.S. Teotia, M. and Singh, R.K. 1981. Hydrogeochemical aspects of endemic skeletal fluorosis in India. An epidemiological study. *Fluoride*, **14**(2): 69-74.

Teotia, S.P.S., Teotia, M. and Singh, D.P. 1985. *Studies in Environmental Science*, Vol **27**. Amsterdam, Elsevier Science Publishers BV, pp 347-355.

Wang, L.F., Huang, J.Z. 1995. *Social Science and Medicine*, 41: 1191-1195.

Whitford, G.M. Fluoride In: Stipanuk, M.H. Editor. Philadelphia WB Saunders Company, 2000. p. 810-822.

Zejmo, M., Stecewicz, I. and Machalinski, B. 1998. *Metabolism Fluoru.*, 8: 161-165.

2013, Impact of Global Climate Change on Earth Ecosystems *Pages* ***131–134***
Editors: **D.R. Khanna, A.K. Chopra, Gagan Matta, Vikas Singh & Rakesh Bhutiani**
Published by: **BIOTECH BOOKS, NEW DELHI**

Chapter 17

Clinicopathologic Study of Subacute Thyroiditis and Hashimoto's Disease

A. Shashi and N. Sharma
Department of Zoology,
Punjabi University, Patiala – 147 002, Punjab

The present investigation evaluate the incidence and clinico-pathological features of different syndromes of thyroiditis. Eighty nine patients with clinical diagnosis of thyroiditis in the age group of 32-56 years were selected from Govt. Rajindra Prasad Medical College and Hospital, Kangra, Himachal Pradesh, India. Sixty five cases of chronic thyroiditis was diagnosed, out of which forty five patients have Hashimoto's thyroiditis. Lymphocytic thyroiditis was diagnosed in twenty four goitrous patients. Subacute granulomatous "De Quervain's" thyroditis was diagnosd in twenty patients. The thyroid function hormones TSH, T3 and T4 of the patients were analyzed by enzyme immune assay technique. All the patients with Hashimoto's and lymphocytic thyroiditis were affected with hypothyroidism, while patients of subacute granulomatous thyroiditis were hyperthyroid. The smears of diagnostic cases revealed presence of scant colloid with diffuse lymphoid infilteration, fibrosis, atrophy of follicular cells and oncocytic changes in some follicular cells in patients of Hashimoto's thyroiditis. Most of the follicular cells formed hurthal cells transformations. Aspirates also exhibited hemosiderin–laden macrophages and multinucleated giant cells with 20-30 nuclei and colloid containing cytoplasm. The FNACs of lymphocytic thyroiditis patients were characterized by presence of numerous lymphoid cells at different stages of maturation, including mature and immature lymphoid cells and a few plasma cells. There were small groups of follicular epithelial cells in aciner or ball like clusters with vasicular nuclei and scant cytoplasm. Aggregates of paravacuolar granulomas with degenerative and proliferative changes of the follicular cells infilterated by lymphocytes were visible in smears of subacute granulomatous thyroiditis. Multinucleated cells with more than 50 nuclei were prominent. In some cases, carrot shaped nuclei were seen in synacial clusters.

Keywords: Hashimoto's Thyroiditis, Lymphocytic thyroiditis, Subacute thyroiditis, TSH, T3, T4.

Introduction

Thyroiditis is an inflammation of the thyroid gland that may be painful and tender when caused by infection, radiation, trauma, or painless when caused by autoimmune conditions, medications, and an idiopathic fibrotic process. The most common forms are Hashimoto's disease, subacute granulomatous thyroiditis, postpartum thyroiditis, subacute lymphocytic thyroiditis, and drug-induced thyroiditis (Bindra and Braunstein, 2006). Subacute thyroiditis is a seasonal disorder that generally affects middle aged women. It is characterized by neck pain and generalized malaise, fatigue, fever and chills following an upper respiratory infection, usually of viral etiology. Fine needle aspiration specimens contain acute inflammatory cells on a background of mixed inflammatory cells and multinucleated giant cells with degenerated thyroid follicular epithelial cells (Thompson, 2002). The present investigation assessed the clinico-pathological features of the different syndromes of thyroiditis in Kangra valley, Himachal Pradesh, India.

Patients and Methods

Eighty nine patients aged 32-56 (mean age 47 ± 11.45 years) affected with clinical thyroiditis, were selected. Serum levels of thyroid stimulating hormone, triiodothyronin and thyroxine were estimated by enzyme immune assay tests on ELISA reader. Preoperative fine needle aspiration cytology was done in all patients. The smears were air dried, fixed in methanol and stained with May- Grunwald Giemsa and Papanicolau stains. The cytopathological results were correlated with clinical features and thyroid function.

Results and Discussion

Eighty nine patients with clinical diagnosis of thyroiditis in the age group of 32-56 years were included. The mean age of patients with subacute thyroiditis was significantly lower than that of those with chronic thyriditis (37.43 ± 4.49 vs 45.49 ± 7.34 years, $p<0.0001$). Sixty five cases of chronic thyroiditis were diagnosed, out of which forty five patients have Hashimoto's thyroiditis. Lymphocytic thyroiditis was diagnosed in twenty four goitrous patients. Subacute granulomatous "De Quervain's" thyroditis was diagnosed in twenty patients. All the patients with Hashimoto's and lymphocytic thyroiditis were affected with hypothyroidism, while patients of subacute granulomatous thyroiditis were hyperthyroid Similar findings were reported by many authors (Peter, 1991; Goudie, 1992) Khlil *et al.* (2004) documented that the mean age of the subacute granulomatous thyroiditis patients was significantly lower than that of those with chronic thyroiditis patients. Patients with chronic thyroiditis revealed subclinical thyroidism, on the other hand a transient elevation in thyroid hormones and reduction in TSH level was noted the subacute granulomatous thyroiditis patients.

Hashimoto's thyroiditis were diagnosed in forty five patients. The smears revealed presence of scant colloid, and abundant cells predominantly lymphoid and epithelial. Synanciums of benign follicular epithelial with nuclear crowding and overlapping were visible. The transformed cells have vasicular nuclei with one or two nucleoli and scant cytoplasm. Most of the follicular cells formed hurthal cells transformations. Monolayered sheets of hurthal cells with granulated cytoplasm were prominent. On high magnification, classic hurthal cells with well defined margins, fine granular eosinophilic cytoplasm and a large, hyperchromatic and pleomorphic nucleus were seen. Aside from lymphoid and epithelial cells, aspirations from thyroiditis patients also revealed presence of hemosederin-ladan macrophages. These were macrophages, which phagocytised ruptured red cells and leads to transformation of hemoglobin into brown colored hemosederin. Smears of a few patients showed presence of multinucleated giant cells with 20-30 nuclei and colloid containing cytoplasm. Nguyun *et al.* (1996) reported presence of numerous lymphoid cells and sheets of hurthal cells with abundant, granular and eosinophilic cytoplasm and slightly or marked nuclear polymorphism in the smears of diagnostic cases. Multinucleated cells and hemosiderin-laden macrophages were noted only in one case. Handa *et al.* (2008) reported lymphocytic infilteration destroying follicular epithelial cells, hurthal cell changes, lymphoid tangles, epithloid cells and multinucleated cells in patients with thyroiditis.

Lymphocytic thyroiditis was diagnosed in twenty six goitrous patients. The FNAC of the lymphocytic thyroiditis patients were moderately cellular smears comprised of benign follicular epithelial cells admixed with an abundant polymorphous benign lymphoid infilterate in a colloidal and hemorrhagic background. Most of the lymphoid cells were lymphocytes and plasmacytes. There were small groups of follicular epithelial cells in aciner or ball like clusters with vasicular nuclei and scant cytoplasm. The follicular cells were varying in number and morphology. Clusters of transformed oncocytic cells surrounded and in filtered by lymphocytes were also visible. The oncocytic cells have abundant eosinophilic and granular cytoplasm and large bizarre nuclei. In a female patient aged 52 years a colloid containing follicle surrounded by a polymorphic lymphoid population was also noted. The colloid material was dense and deeply eosinophilic. Cabay *et al.* (2008) noted a few groups of benign follicular cells scattered within an filteration of mature lymphocytes in smear of the aspiraes of a 21 year old woman presented with chronic lymphocytic thyroiditis.

Subacute granulomatous thyroiditis was found in twenty patients. The smears were highly cellular with scant colloid material in the background. The cell population mostly contains nacrotic follicular cells, neutrophils, fibroblasts, lymphocytes, macrophages and multinucleated giant cells. Aggregates of paravacuolar granules with degenerative and proliferative changes of the follicular cells infiltered by lymphocytes were visible in FNAC smears of almost all the patients with sub acute thyroiditis. Presence of multinucleated giant cells was a key feature of subacute thyroiditis. Multinucleated cells with more than 50 nuclei were present in almost all patients. In some cases carrot shaped nuclei were seen in synacial clusters. Shabb and Salti (2006) documented salient cytologic features included cellular smears, multinucleated giant cells in 100 per cent of cases, some ingesting colloid or neutrophils, fibrous fragments with enmeshed inflammatory cells were a constant feature; follicular cells were scant to absent in most cases. Granulomas were rare. Colloid, when present was thick, with central cracks and frayed edges.

References

Bindra A, Braunstein GD. *Am Fam Physician* 2006; 73(10):1769-1776.

Cabay RJ, Salem F. *Diag Cytopathol* 2008; 37(3):191.

Goudie RB 1992. The thyroid gland. In: *Oxford Textbook of Patholgy*. Janes OD, Isaacson PG, Wright NA, Dick HM, Slack, MP, eds. 26, pp 1940-1959.

Handa U, Garg S, Mohan H, Nagarkar N. *J Cytol.*, 2008; 25(1): 13-17.

Khalil MR, Hamza MA, Moussa M, *et al.*, *Egyptian J Surg* 2004; 23(2):126-132.

Nguyen GK, Ginsberg J, Crockford PM, *et al.*, *Diag Cytopathol* 1996; 16(6):531-536.

Peter AS. *Med. Clin North Am* 1991; 75:61-77.

Shabb NS, Salti IS. *Diag cytopathol* 2006; 34(1):18-23.

Thompson LDR, Heffess CS. *ENT J* 2002; 9.

2013, Impact of Global Climate Change on Earth Ecosystems *Pages 135–140*
Editors: **D.R. Khanna, A.K. Chopra, Gagan Matta, Vikas Singh & Rakesh Bhutiani**
Published by: **BIOTECH BOOKS, NEW DELHI**

Chapter 18
Prevalence and Clinical Aspects of Thyroid Disorders in Himachal Pradesh, India

A. Shashi and N. Sharma
Department of Zoology,
Punjabi University, Patiala – 147 002, Punjab

Diseases of the thyroid are manifested by alterations in hormone secretion, enlargement of the thyroid gland (goitre) or both. The principal diseases of the thyroid are goitre (diffuse or nodular), hypo or hyperthyroidism, thyroditis and neoplasm. The simple goitre is extremely common throughout the world and is most prevalent in mountainous areas. Thyroid diseases are multifactorial with contributions from genetic and environmental factors. The present study describe the incidence of various types of thyroid disorders in eight hundred twelve patients in the age group of 25-55 years (mean age - 47)selected from different areas of Himachal Pradesh, India. Metabolic lesions were found in six hundred fifty six patients (80.7 per cent), non-neoplastic lasions in 111 (13.7 per cent) and neoplastic lesions in forty five (5.54 per cent) thyroid patients. Among patients with metabolic lesions, three hundred sixty six patients (55.7 per cent) suffered from hypothyroidism, while three hundred (45.7 per cent) were affected with hyperthyroidism. In non neoplastic lesions, thyroiditis was most prevalent and was found in eighty nine (80.2 per cent) patients, colloid goitre in sixteen (14.5 per cent) and multinodular goitre in six (5.4 per cent) patients. In neoplastic lesions, papillary carcinoma was most frequent (51.2 per cent) cancer seen in series, followed by follicular carcinoma (26.7 per cent) and medullary carcinoma (22.3 per cent). Our results indicate that thyroid disorders are quite prevalent in Himachal Pradesh, India. Goiterogenic and antithyroidal substances present in food and water in environment other than iodine deficience may have role for the persistence of endemic goitre inspite of adequate iodine intake.

Keywords: *Carcinoma, Goitre, Metabolic lesions, Risk factors, Thyroid disorders.*

Introduction

Thyroid hormones are essential in all phases of life cycle, including foetal and neonatal neurological development, overall growth, development, physical and mental efficiency, energy production and reproduction. Diseases of thyroid gland are manifested by alterations in hormone secretion, enlargement of thyroid gland (goiter) or both (Wortofsky, 1998). The principal diseases of thyroid gland are goiter (diffuse or nodular), hypothyroidism, hyperthyroidism, thyroiditis and neoplasms. The simple goitre is extremely common throughout the world and is most prevalent in mountainous areas. Thyroid diseases are multifactorial with contributions from genetic and environmental factors (Tsegaye and Ergete, 2003). Goiterogenic and antithyroidal substances present in food and water in environment other than iodine deficiency may have role for the persistence of endemic goiter inspite of adequate iodine intake (Chandra, 2011). The present study describes the incidence of various types of thyroid disorders after iodine supplementation programme in Kangra Valley, Himachal Pradesh, India.

Materials and Methods

The present study was conducted on 812 patients in the age group of 25-65 (Mean age- 47± 11.54) suffering from various thyroid disorders in Kangra valley, Himachal Pradesh, India. All patients were subjected to thorough history taking and complete clinical examination with special emphasis on manifestations of hypo- or hyperthyroidism and clinical criteria of the goiter and its association with pain. Serum levels of thyroid stimulating hormone, triiodothyronin and thyroxine were estimated by enzyme immune assay tests on ELISA reader. Results were presented as Mean ± SD. Measures of significance between groups were calculated using the χ^2 test and analysis of varience (ANOVA) by using SPSS (19.0) statistics package. The symptom index was calculated in the manner of Billewicz *et al.* (1969) and. P value less than 0.001 was considered significant.

Results

Eight hundred twelve patients in the age group of 25-65 years (mean age – 47 ± 11.54 years) selected from Kangra Valley of Himachal Pradesh, India. The pattern of thyroid lesions in studied population was presented in Table 18.1. Metabolic lesions were found in six hundred fifty six patients (80.7 per cent), non-neoplastic lasions in 111 (13.7 per cent) and neoplastic lesions in forty five (5.54 per cent) thyroid patients. Among patients with metabolic lesions, three hundred sixty six patients (55.7 per cent) suffered from hypothyroidism, while three hundred (45.7 per cent) were affected with hyperthyroidism. In non neoplastic lesions, thyroiditis was most prevalent and was found in eighty nine (80.2 per cent) patients, colloid goitre in sixteen (14.5 per cent) and multinodular goitre in six (5.4 per cent) patients. In neoplastic lesions, papillary carcinoma was most frequent (51.2 per cent) cancer seen in series, followed by follicular carcinoma (26.7 per cent) and medullary carcinoma (22.3 per cent).. In contrast to our results Tsegaye and Ergete (2003) reported nodular colloid goiter in 76.9 per cent cases, adenoma in 12.8 per cent, thyroiditis in 8.2 per cent and carcinoma in 2.1 per cent cases in a retrospective analysis.

Table 18.1: Pattern of Thyroid Lesions in the Study Population

Sl.No.	*Lesion*	*Number of Patients*	*Percentage of Sample Size*
I.	Metabolic lesions (656)		
	✩ Hypothyroidism	366	55.79
	✩ Hyperthyroidism	290	44.21
2.	Non-Neoplastic Lesion	(111)	
	Hashmoto's Thyroiditis	45	40.54
	Lymphocytic Thyroiditis	24	21.62
	Subacute Thyroiditis	20	18.01
	Colloid Goitre	16	14.41
	Multinodular Goitre	6	5.42
3.	Neoplastic Lesions (45)		
	Papillary Carcinoma	23	51.11
	Follicular Carcinoma	12	26.67
	Medullary Carcinoma	10	22.22

The age distribution of the specific thyroid disease entities as shown in Table 18.2. Maximum number of patient were in the age group of 45-55 in all type of thyroid disorders. There was a significant trend seen in number of patients from younger to older age group, while last decade showed less number of patients in comparison to 45-55. This was in agreement with some previous studies (Mengistu, 1992; Mekones, 1996). Bjoro *et al.* (2000) reported hyperthyroidism was 2.5 per cent in females and 0.6 per cent in males, hypothyroidism 4.8 per cent and 0.9 per cent, and goitre 2.9 per cent and 0.4 per cent respectively. A report from Colorado stated a prevalence of elevated TSH (> 5.1 mU/l) of 9.5 per cent in the adult population increasing with age from 3 to 16 per cent in males and from 4 to 21 per cent in females (Canaris *et al.*, 2000). In contrast, Uzunlulu *et al.* (2007) reported that only female gender was associated with presence of thyroid dysfunction. Ahmad *et al.* (2009) noted that age, gender, race and area all have an appreciable effect on the levels T4, T3 and TSH.

The thyroid status of the studied patients was summarized in Table 18.3. 5.78 per cent patients were euthyroids, while abnormal thyroid status was observed in 94.22 per cent patients. There were 56.03 per cent patients with an elevated TSH concentration, most of whom were overt hypothyroid. Among those with an elevated serum TSH concentration, 207 subjects had a level between 4.5 – 10.0 μIU/ml, while

248 subjects had a value greater than 10 µIU/ml. 38.19 per cent patients were affected with hyperthyroidism. Most of the hyperthyroid patients revealed TSH concentration between 0.10 – 0.39 µIU/ml. only 56 patients had a tsh concentration below > 0.10 µIU/ml.

Table 18.2: Distribution of Different Thyroid Lesions in Relation with Age

Age (Years)	*Metabolic Lesions*	*Non-Neoplastic Lesions*	*Neoplastic Lesions*
25-35	134	19	08
35-45	152	26	11
45-55	193	39	17
55-65	177	27	9
Total	656	111	45

Table 18.3: Prevalence of Thyroid Abnormalities in Studied Population

Thyroid Status	*Number of Patients*	*Percentage of Sample Size*
Euthyroid	47	5.78
Subclinical Hypothyroid	146	17.98
Overt Hypothyroid	309	38.05
Subclinical Hyperthyroid	110	13.55
Overt Hyperthyroid	200	24.64

The frequencies of clinical symptoms among patients of overt and subclinical hypothyroid were shown in Table 18.4. Overt hypothyroidism patients reported a greater percentage of symptoms than did the subclinical hypothyroid group. Most frequent symptoms among overt hypothyroidism patients were weight gain (60.11 per cent), bradykienesis (60.11 per cent), weakness (59.11 per cent), decreased appetite (57.37 per cent), constipation (50.54 per cent) and intolrence to cold. However, some symptoms were reported in very low percentage among subclinicalgroup, such as cold skin (6.28 per cent), paresthesias (6.28 per cent) and depression (6.83 per cent). Univariate analysis showed that most of the clinical symptoms differ significantly ($p<0.001$) between overt hypothyroid and subclinical hypothyroid patients, only two symptoms cold skin and diminished sweating showed non- significant difference.

The clinical index for the diagnosis of hyperthyroidism was given in Table 18.5. There was a positive association between proportion of symptoms reported and progressive thyroid failure. Overt hyperthyroidism patients reported a greater percentage of symptoms than did the subclinical hyperthyroid group. Most frequent symptoms among overt hyperthyroidism patients were weight gain (58.06 per cent), Apatite decrease (58.06 per cent), fatigue (53.22 per cent), insomnia (47.41 per cent), tremor (46.12 per cent) and intolrence to heat. Univariate analysis showed that all the symptoms differ significantly ($p<0.001$) between overt hyperthyroid and subclinical hyperthyroid patients.

Table 18.4: Frequencies of Clinical Symptoms in Overt and Subclinical Hypothyroid Patients

Symptoms	*Overt Hyperthyroidism Percentage*	*Subclinical Hyperthyroidism Percentage*	*95 per cent Confidence Interval*	*Odd's Ratio*
Weight gain	60.11 (220)	25.13 (92)	–	–
Bradykienesis	60.11 (220)	12.29 (45)	–	–
Weakness	59.11 (216)	18.31 (67)	22.48–180.36	63.37
Decreased appatite	57.37 (210)	26.78 (98)	4.99–21.18	10.28
Constipation	50.54 (185)	20.76 (76)	2.99–7.91	4.87
Intolerance to cold	48.91 (168)	17.75 (65)	2.56–6.32	4.02
Alopecia	45.91 (168)	8.74 (32)	6.98–18.99	11.51
Depression	36.88 (135)	6.83 (25)	4.62–12.79	7.69
Coarse skin	28.68 (105)	19.94 (78)	0.53–1.21	0.79
Periorbital puffiness	25.68 (94)	9.56 (35)	1.48–3.76	2.37
Neck enlargement	24.31 (89)	2.12 (12)	3.96–14.52	7.58
Paresthesias	19.67 (72)	6.28 (23)	1.54–4.41	2.61
Diminished sweating	17.75 (65)	9.56 (35)	0.82–2.14	1.33
Cold skin	17.21 (63)	6.28 (23)	1.26–3.66	2.15

Table 18.5: Frequencies of Clinical Symptoms in Overt and Subclinical Hyperthyroid Patients

Symptoms	*Overt Hyperthyroidism Percentage*	*Subclinical Hyperthyroidism Percentage*	*95 per cent Confidence Interval*	*Odd's Ratio*
Weight loss	58.06 (180)	23.87 (74)	–	–
Appatite decreased	58.06 (180)	20.96 (65)	–	–
Fatigue	53.22 (165)	20.32 (63)	4.28–15.71	8.21
Insomnia	47.41 (147)	19.35 (60)	2.17–6.32	3.71
Tremor	46.12 (143)	17.41 (54)	2.38–6.74	4.01
Heat intolerance	45.16 (140)	15.81 (49)	2.60–7.29	4.35
Palpitations	43.54 (135)	14.51 (45)	2.61–7.20	4.22
Skin changes	37.06 (115)	12.91 (40)	1.89–5.07	3.09
Staring gaze	37.09 (115)	11.93 (37)	2.12–5.75	3.49
Exophthalmos	36.12 (112)	10.32 (32)	2.41–6.69	4.01
Alopecia	26.77 (83)	9.03 (28)	1.49–4.21	2.51
Trembling hands	13.84 (54)	5.89 (23)	0.93–2.84	1.62
Thyroid enlargement	9.48 (37)	5.89 (23)	0.55–1.76	0.98

More symptoms were observed in overt cases in comparison to subclinical and euthyroid subjects in this study. There was a positive association between the proportion of symptoms reported and progressive thyroid failure. Several investigators support the usefulness of multiple symptoms as a diagnostic tool for thyroid dysfunction (Helfand and Crapo, 1990; Canaris *et al.*, 1997; Zulewski *et al.*, 1997). Canaris *et al.* (2000) reported more symptoms in hypothyroid patients in comparison to euthyroid individuals. Individual symptom sensitivities were low and there was a weak relationship between symptoms reported and thyroid failure.

References

Canaris GJ, Manowitz NR, Mayor G, Ridgeway EC. Arch Intern Med 2000; 160: 526-34.

Canaris GJ, Steiner JF, Ridgeway EC. J Gen Intern Med 1997; 12: 544-50.

Chandra, A.K. Al Ameen J Med Sci 2011. 213- 4.

Helfand M, Crapo LM. Screening for thyroid disease. Ann Intern Med 1990; 112: 840-9.

Mekones E.. East Afr Med J 1996; 73: 264-7.

Mengistu M. East Afr Med J 1992; 69: 515-9.

Tsegaye B, Ergete W. East Afr Med J 2003; 80(10): 525-8.

Uzunlulu M, Yorulmaz E, Oguz A. Endocine J 2007; 54(1):71-6.

Zulewski H, Muller B, Exer P, Miserez AR, Staub JJ. J Clin Endocrinol Metabol 1997; 82: 771-6.

2013, Impact of Global Climate Change on Earth Ecosystems *Pages* **141–147**
Editors: **D.R. Khanna, A.K. Chopra, Gagan Matta, Vikas Singh & Rakesh Bhutiani**
Published by: **BIOTECH BOOKS, NEW DELHI**

Chapter 19

Conservation of Water by Development of Drought Tolerant Lines in Rice (*Oryza sativa* L.) through *In vitro* Culture

***Dipti Verma*[1], *Rashmi Verma*[1] *and Alok Shukla*[2]**
[1]*Department of Plant Physiology,*
College of Basic Sciences and Humanities,
G.B. Pant University of Agriculture and Technology,
Pantnagar – 263 145, U.S. Nagar, Uttarakhand
[2]*Department of Biotechnology, Kumaun University,*
Bhimtal – 263 136, Nainitaal, Uttarakhand

To mitigate the increasing water scarcity in Asia, it is necessary to develop alternate ways of growing rice that use less water while maintaining high yields. In the present study an attempt was made for the development of somaclones suitable for drought conditions through *in vitro* culture. Molecules of PEG (polyethylene glycol mwt.6000), were used as they are drought inducer. Mature embryos of four rice cultivars namely Govind, Jaya, Pusa Basmati-1 and PR113 were taken as explants. Selection was made on different concentration of PEG in MS media. Osmotic stressed calli were selected on the basis of physical appearance, relative growth rate, water, proline and MDA content. It was observed that rice cultivars Govind and PR113 performed best during drought stress. These *in vitro*

* Corresponding Author: E-mail: dipti82verma@gmail.com

developed somaclones can further be used as a model system for developments of drought tolerant lines so that, tremendous amount of water can be saved by growing rice in terms of productive and sustainable agriculture.

Keywords: *Water, Drought, Rice, Somaclones, PEG (polyethylene glycol).*

Introduction

Rice is the world's single most important food crop and a primary food for more than a third of the world's population. Under both natural and agriculture conditions plants are frequently exposed to stresses. Several biotic and abiotic stress factors adversely affect cultivation of rice crop in various geographical locales. The abiotic stress results in more damage to growth and yield of rice compared to the biotic stresses. Drought stress is the major constraint to rice production and yield stability. Unreliable monsoon and uneven distribution of rainfall cause year-to-year fluctuations in crop yields Irrigated rice requires lot of water, about 3000-5000 liter of water is used to produce 1kg of grain. Water is a looming crisis due to competition among agricultural, industrial, environmental and domestic users. By 2025, 30 per cent of the human population would be threatened by water scarcity because worldwide, 85 per cent of the total available water is used by agriculture sector alone of which 75 per cent is used for rice cultivation. To mitigate the increasing water scarcity in Asia, it is necessary to develop alternate ways of growing rice that use less water while maintaining high yields. Development of drought-tolerant lines, through the use of biotechnological approaches such as *in vitro* screening, would be appropriate as *in vitro* culture provides a uniform population of synchronously developing plant cells without involving regulatory mechanisms that naturally repaired at the whole plant level. In the present study an attempt was made for the development of somaclones suitable for drought conditions through *in vitro* culture.

Materials and Methods

Seeds of rice cultivar namely Govind, Jaya, Pusa Basmati-1 and PR113 were taken as explants for callus induction and regeneration on MS media supplemented with different concentration of phytohormones. Seeds were obtained from Seed Production Centre, of G.B.Pant University of Agriculture and Technology Pantnagar, U.S. Nagar (Uttarakhand). Sterilization of seeds of rice cultivar was performed by treating them with 70 per cent ethanol followed by treatment with 1 per cent sodium hypochlorite solution, then treated with 0.5 per cent mercuric chloride. Seeds of rice cultivar were placed on agar solidified MS media containing 3.0 per cent sucrose. Growth regulators used in the study were: 2, 4-D, NAA, BAP and Kn. Molecules of PEG (polyethylene glycol mwt.6000), were used as they are drought inducer and small enough to influence the osmotic potential, in cultured cells. Selection was made on different concentration of PEG in MS media as 30g/l, 50g/l and 70g/l. Plants developed through tissue culture were hardened and after 15-20 days, for proper growth they were transferred in the field at Dr.N.E. Borolaug Crop Research Centre of G.B.P.U.A. &T., Pantnagar located in tarai plains about 30 Km southwards of foothill of Shivalik range of Himalaya at 29° N latitude, 79° 29' E longitudes and at an altitude of 243.8 meters above the mean sea level.

Results

Callus Induction Frequency

The data was statistically significant for callus induction frequency of all the rice cultivars namely Govind, Jaya, Pusa Basmati-1 and PR113. The highest callus induction frequency was shown by Jaya (62.33 per cent) followed by Pusa Basmati-1 (58.00 per cent), PR113 (53.58 per cent) and Govind recorded the lowest callus induction frequency (45.67 per cent)

Table 19.1: Induction Frequency of Callus from Mature Embryo in Rice Cultivars Govind, Jaya, PB-1 and PR 113 on MS Media Supplemented with 13.5mM 2, 4-D

Cultivars	*Induction Frequency (per cent)*
Govind	45.67
Jaya	62.33
PB-1	58.00
PR 113	53.58
S.Em.±	1.58
CD at 5 per cent	4.65

Effect of PEG on Callus Proliferation

Proliferation of the callus was measured on the basis of change in volume in MS media supplemented with varying concentration of PEG from 30, 50, and 70 g/l. In all the rice cultivars, as compared to control the proliferating capacity decreased with the increase in the concentration of the PEG in the MS media.

Table 19.2: Volume of Calli (cm³) Transferred to MS Medium Supplemented with Varying Concentration of PEG

Cultivars	*PEG Concentration (g/l)*			
	0	*30*	*50*	*70*
Govind	3.82±0.02	3.60±0.03	3.59±0.03	3.18±0.03
Jaya	3.49±0.07	3.21±0.05	3.04±0.04	2.99±0.01
PB-1	2.63±0.05	2.40±0.08	2.09±0.06	1.97±0.01
PR 113	3.52±0.02	3.05±0.02	2.72±0.02	2.42±0.03
Mean	3.37	3.07	2.86	2.64
S.Em.±		0.0216		
CD at 5 per cent		0.0619		

Effect of PEG on the Callus Growth

The effect of stress on callus proliferation of rice cultivars namely Govind, Jaya, Pusa Basmati-1, and PR113 was examined. On comparison with control, decrease in fresh weight was observed with increase in concentration of PEG from 50 to 350 g/l in the MS media. The maximum growth of callus was observed in rice cultivar Govind followed by PR113, Jaya, and Pusa Basmati-1.

Table 19.3: Effect on Relative Growth Rate (RGR) of Callus of Different Rice Cultivars Stressed with Different Concentration of PEG

Cultivars	*PEG Concentration (g/l)*			
	0	*30*	*50*	*70*
Govind	384.22 ±7.46	373.67 ±4.56	375.55±9.46	369.67±4.56
Jaya	271.33 ±6.12	271.67 ±2.98	275.99 ±8.67	191.67 ±3.25
PB-1	175.33±7.35	176.00 ±2.65	160.22±6.45	149.00±2.78
PR 113	357.55±8.42	314.67±4.23	311.99±10.23	236.67 ±2.87
Mean	297.11	284.02	280.94	237.75
S.Em.±			5.43	
CD at 5 per cent			16.26	

Proline content of PEG stressed callus

The total free proline content (μg/g fresh weight) in callus of different rice cultivars namely Govind, Jaya, Pusa Basmati-1 and PR113, stressed with different concentration of PEG in MS media was examined. Proline content was observed to increase with increase in PEG concentration in MS media.

Table 19.4: Total Free Proline Content (mg g^{-1} fr. wt.) in Callus of Different Varieties of Rice Callus Stressed with Different Concentrations of PEG

Cultivars	*PEG Concentration (g/l)*			
	0	*30*	*50*	*70*
Govind	103.22 ±1.22	149.11 ±7.65	161.32±2.13	164.44 ±7.45
Jaya	106.33 ±0.99	109.78±3.25	125.35±1.45	149.11±7.65
PB-1	92.33±1.02	105.77 ±7.65	127.43±1.32	135.55±9.46
PR 113	108.00±1.11	127.77±1.23	131.00 ±1.45	160.22±6.45
Mean	102.46	123.11	136.28	152.33
S.Em.±			0.32	
CD at 5 per cent			0.96	

Effect of PEG on Concentration of MDA (m mol/g fr. wt.)

MDA content (m mol/g fr. wt.) in callus of rice cultivars namely Govind, Jaya, Pusa Basmati-1 and PR113 during PEG stress was represented in Table 5. The data was statistically significant for cultivars as well as for different concentration of PEG in the MS media. The maximum increase in MDA content was observed for rice cultivar Govind.

Table 19.5: MDA Content (m mol g^{-1} fr.wt.) in the Callus of Different Rice Varieties Stressed with Different Concentrations of PEG on MS Media

Cultivars	*PEG Concentration (g/l)*			
	0	*30*	*50*	*70*
Govind	0.51 ±0.11	2.26 ±0.03	3.22±0.09	3.75 ±0.11
Jaya	1.31 ±0.09	1.34 ±0.04	1.37 ±0.08	2.15 ±0.08
PB-1	1.42 ±0.12	1.47 ±0.23	1.54 ±0.11	2.63 ±0.09
PR 113	0.46 ±0.02	1.04 ±0.01	1.64 ±0.08	8.92±0.24
Mean	0.93	1.53	1.94	4.36
S.Em.±		0.054		
CD at 5 per cent		0.154		

Discussion

Callus induction is one of the substantial step for selecting the suitability of genotypes for tissue-culture-based research and for plant improvement (Carsono and Yoshida, 2006). In the present study, MS media supplemented with 13.5mM 2, 4-D was found optimum for callus induction in all rice cultivars namely Govind, Jaya, Pusa Basmati-1 and PR113. These results are supported by the fact that appearance time, frequency, and the phenotype of the calli varied among the rice cultivars due to genotypic differences or due to different endogenous level of phytohormones (Lee *et al.*, 2002; Hoque and Mansfield, 2004).

In present study, fresh weight and relative growth rate of callus was observed to decrease with increase in PEG concentration from 30g/l to 70g/l in MS media in rice cultivars namely Govind, Jaya, Pusa Basmati-1and PR113. Similar effects of drought-inducers (PEG) on callus production and regeneration were observed in several plant species (Adkins *et al.*, 1995; Mercado *et al.*, 2000; Taghian, 2002). Dehydration induced by PEG reduces the availability of water and thus turgidity and growth (Heyser and Nabors, 1981).

This increase in proline content was found to be highest in the rice cultivar Govind, followed by PR113, Jaya and Pusa Basmati-1. Similar results are reported earlier in PEG-treated wheat callus with the increasing PEG concentration in medium (Bajji *et al.*, 2000). It is evident that iso-osmotic stress developed by PEG significantly increased production of proline (Al-Khayri 2002; Al-Khayri and Al-Bahrany; 2002).

In present study, malondialdehyde (MDA) content was observed to increase with increase in PEG concentration. Maximum increase in malondialdehyde (MDA) content was observed in rice cultivar Govind, followed by Pusa Basmati-1, Jaya and PR113. Increase in malondialdehyde (MDA) content under PEG stress, may be a protective feature against membrane lipid peroxidation. The above result is consistent with the findings that, enhanced MDA content is common in plant cells subjected to water stress (Sairam *et al.*, 1997; Sairam *et al.*, 1998; Zhang and Kirkham, 1996).

Conclusion

The results reported here suggest that the gradual selection is the most efficient procedure for the establishment of drought-tolerant rice cell lines. It was observed that rice cultivars Govind and PR113 performed best during drought stress. These *in vitro* developed somaclones can further be used as a model system for developments of drought tolerant lines in other cultivars of rice so that, tremendous amount of water can be saved by growing rice in terms of productive and sustainable agriculture.

Acknowledgements

This research was supported by the grant received through All India Coordinated Rice Improvement Programme of ICAR, New Delhi, India. The authors wish to acknowledge Dean, College of Basic Sciences and Humanities, and Director, Experiment station of G. B. Pant University of Agriculture and Technology, Pantnagar-263145, India for providing all necessary facilities for the present investigation.

References

Adkins, S.W., Kunanuvatchaidach, R. and Godwin, I. D. 1995. Somaclonal variation in rice: Drought tolerance and other agronomic characters. *Australian J. of Botany.* 43: 201-209.

Al-Khayri, J.M. 2002. Growth, proline accumulation, and ion content in sodium chloride-stressed callus of date palm. *In Vitro Cell Develop. Biol Plant.* 38: 79–82.

Al-Khayri, J.M. and A.M. Al-Bahrany, 2002. Callus growth and proline accumulation in response to sorbitol and sucrose-induced osmotic stress in rice (*Oryza sativa* L.). *Biol. Plant.* 45: 609-611.

Bajji, M., Lutts, S. and Kinet, J.M. 2000. Physiological changes after exposure to and recovery from polyethylene glycol-induced water deficit in callus cultures issued from durum wheat (*Triticum durum* Desf.) cultivars differing in drought resistance. *J. Plant. Physiol.* 156: 75-83.

Carsono, N. and Yoshida, T., 2006. Identification of Callus Induction Potential of 15 Indonesian Rice Genotypes. *Plant Prod. Sci.* 9: 65-70.

Heyser, J.W., and Nabors, M.W. 1981. Osmotic adjustment of cultures tobacco cells (*Nicotiana tabacum* var. Samsum) grown on sodium chloride. *Plant Physiol.* 67: 720-727.

Hoque, Md.E. and Mansfield, J.W. 2004. Effect of genotype and explant age on callus induction and subsequent plant regeneration from root-derived callus of Indica rice genotypes. *Plant Cell Tiss. Org. Cult.* 78: 217-223.

Lee, K., Jeon, H. and Kim, M. 2002. Optimization of mature embryo-based in vitro culture system for high-frequency somatic embryogenic callus induction and plant regeneration from Japonica rice cultivars. *Plant Cell Tiss. Org. Cult.* 71: 237-244.

Mercado, J.A.; Sancho, C.; Jimenez, B.S.; Peran, U.R.; Pliego, A.F. and Quesada, M.A. 2000. Assessment of *in vitro* growth of apical stem sections and adventitious organogenesis to evaluate salinity tolerance in cultivated tomato. *Pl. Cell Tissue and Organ Culture.* 62: 101-106.

Sairam, R.K., Deshmukh, P.S. and Shukla, D.S. 1997. Tolerance of drought and temperature stress in relation to increased antioxidant enzyme activity in wheat. *Journal of Agronomy and Crop Science.* 178: 171–178.

Sairam, R.K., Shukla, D.S. and Saxena, D.C. 1998. Stress-induced injury and antioxidant enzymes in relation to drought tolerance in wheat genotypes. Biol. Plant. 40: 357-364.

Taghian, A.S. 2002. Expression of Mutual Tolerance to Drought and Salt Stresses of the in vitro Selected Clones of Sugarcane. Proceeding of the 3rd Scientific Conference of Agriculture Sciences, Fac. Of Agric. Assiut Univ. Egypt October. pp. 441-262.

Zhang, J., Kirkham, M.B. 1996. Antioxidant responses to drought in sunflower and sorghum seedlings. *New Phytol.* 132: 361- 373.

2013, Impact of Global Climate Change on Earth Ecosystems *Pages* **149–153**
Editors: **D.R. Khanna, A.K. Chopra, Gagan Matta, Vikas Singh & Rakesh Bhutiani**
Published by: **BIOTECH BOOKS, NEW DELHI**

Chapter 20

A Study on Coastlines and Bio-Shields

B. Deoli Kanchan, Debajeet Sarma, Ranganath Bhargav, Sikkewal Rohit, Khanna Bharat and Acharya Saransh
UPES

Coastal environments are uniquely dynamic ecosystems, often negotiating between sandy stretches of land and vagaries of the ocean. Over the last century, in the absence of regulations, these fragile fringes of land have experienced rapid urbanization and infrastructure development rendering them more vulnerable to man-made disasters. To confront with the prospect of rising sea levels, and frequency of natural calamities, the immediacy of conserving the unique coastal biodiversity of India has become imperative. The conservation of the coastal biodiversities however, has been one of the most contentious debates of the 21st century. Firstly, there is a conflict between the construction activity on the coast and environmentalists legitimately striving for its conservation. Secondly there is a complex debate of equitable distribution of benefits and economic gains derived from natural resources against corporate interest to privatize natural reserves. So keeping the integrity of settlements and biodiversity, this paper emphasizes on enhancing natural vegetation along the coastline which can act as a shield to various natural calamities.

Introduction

Bio-shields are vegetative shelter belts which act as a barrier against natural calamities such as tsunami. Bio-shields became a hot topic in the wake of the December

2004 tsunami which wreaked havoc in Southern India and South-East Asia. The concept of bioshields gained even more support after Hurricane Katrina hit the USA coas in 2005, with many stories in the press and primary literature viewing it as a policy-focusing event. In May 2008, a Category Four cyclone, Cyclone Nargis, struck Myanmar (Burma) causing over 100,000 fatalities (Rodriguez *et al.*, 2009). While damage from the 200 km/hr winds and rain was extensive, the 4 m storm surge inundated large areas of low-lying country. Particular attention has been given to mangroves and more recently, Casurina Equisetfolia which has been introduced in the Indian Ocean and Carribean Sea Regions.

As part of an ongoing study and survey undertaken by the Ashoka Trust for Research in Ecology and the Environment, of the 20 odd villages surveyed in Kariakal and Nagapattinam districts, only two villages actually had 'bioshields', *i.e.*, casuarina plantations, in front of the village. These happened to be small villages with very few active fishermen and boats. The rest of the 'bioshields' are being planted in locations that might be adjacent to the villages or behind them.

The model currently being followed in the implementation of the World Bank's shelterbelt programme is a mixture of the joint forest management (JFM) and social forestry programmes of the Forest Department. The Forest Department in each hamlet does a PRA based on which a village level micro-plan is made and Village Forest Committees (VFCs) are formed. This does lay some partial emphasis on process and participation, but the ground reality, the quality of process and participation is not known. As part of the programme, a number of entry point activities are executed. These are basically to build the trust of the community and get their cooperation. However, unlike other areas, coastal communities are unfamiliar with exercises and programmes such as JFM, social forestry and village micro-plans.

Sand Dunes as Bioshields

Sand dunes appear to provide the most immediate form of coastal protection. In the wake of tsunami, extensive dune plantations have been undertaken and the construction of permanent settlements, and sea-walls and has led to the disappearance of beaches, such as in Puducherry. Dunes act as a shield against storm surges and tsunamis. Many coastal settlements built behind or on coastal dune formations were protected from the tsunami. Unfortunately, dunes are not considered a worthy ecosystem, sometimes condemned even by ecological restoration

projects. However, they are integral to the livelihood of fishing communities. In the recent years, their population is decreasing due to planting of exotic plants such as casuarinas for the mitigation of natural disasters

Review of Literature

Feagin, R.A., Mukherjee, N., Shanker, K. *et al.* (2010) studied about Bioshields taking into consideration the pros and cons and concluded that bioshields are effective against natural calamities only in the short run; as they will invade native ecosystems such as mangroves and destroy natural habitat. Both of them,along with, Nibedita Mukherjee,, Andrew H. Baird, Joshua Cinner, Alexander M. Kerr, Nico Koedam, Aarthi Sridhar, Rohan Arthur, L.P. Jayatissa, Danny Lo Seen, Manju Menon, Sudarshan Rodriguez, Md. Shamsuddoha and Farid DahdouhGuebas(2009) analysed data on the natural disasters since 2004; and also on bio-shields and came to the following conclusions:

1. *Casuarina* plantations may negatively affect turtle nesting areas
2. Mangroves can protect but they do not grow everywhere
3. There is nothing wrong in restoring mangroves where they have been previously destroyed
4. The location and type of bioshield to be used has to be investigated

R.S. Bhalla (2007), studied on large scale bio-shield plantations and came to the premise that These plantations may obliterate the natural sand dune ecosystems which are an important natural defence and provide a range of ecological goods and services.

Durganand Balsavar (2011) studied the coastlines in various parts of South India such as Poonthura(near Thiruvananthapuram) and Tuticorin Roche Park Region and Cuddalore and found out that

1. In Poonthura, the damage has been reduced by using a groyne(a wall or jetty built out from a riverbank or seashore to control erosion)
2. In Tuticorin, mangrove vegetation has helped mitigate the force of waves to some extent.

He has concluded that selection of plant species for coastal vegetation requires prudent planning taking into account their intrinsic links to conserving interdependent features of the ecosystems and livelihood of the communities.

F. Dahdouh Guebas(2005) along with some Sri-Lankan Scientists studied mangrove vegetations along the Sri Lankan Coast and concluded that mangrove vegetation functions as a physical barrier against tidal and ocean influences due to their large above ground aerial root systems and standing crop.

V. Selvam,T. Ravishankar,V.M. Karunagaran,R. Ramasubramanian,P. Eganathan,A. K. Parida(2005) studied coastal habitats and concluded how crucial a role was played by mangrove forests and associated wetlands in mitigating the impact of tsunami.

Objectives

1. To study status of coastlines and bioshields in India
2. To study the impact of bioshields on the coastlines.

Results and Discussions

Villages like T.S.Pettai, Vadakku Pichavaram, Killai Fisher Colony, MGR Nagar and Kalaingar Nagar which were just 1-2 km from the seashore were not affected to that extent by the tsunami as they were under the direct physical coverage of the mangrove wetlands.This just showed the importance of mangrove plantations in mitigating the effect of natural disasters. In areas where the mangrove stands were totally or partially cleared in this district, the damage was high. Mangrove plants posses a number of unique adaptive features such as:

- ✰ Very low oxygen in mangrove soil
- ✰ Pores in the roots through which gaseous exchange takes place
- ✰ Tolerance to soil salinity by salt excretion and salt exclusion.
- ✰ Animals native to the mangrove habitat are crabs, salt water crocodiles and sea-otters. Crabs of the mangrove environment are called ecosystem engineers since they facilitate air circulation in the soil and thereby influence growth and productivity of the mangrove trees.
- ✰ Along much of the world's coastline, exotic plants such as Casuarina have been introduced for the purpouses of reducing the impact of natural disturbances; but they have only short term effects at the expense of long term ecological sustainability.
- ✰ Species such as *Rhizophora* spp., *Bruguiera* spp. *Ceriops* spp. are planted in the low tidal zone while *Avicennia officina lis* is planted just a few feet away from the low tidal zone whereas high saline tolerant *Avicennia marina* is planted both in mid-tidal and high tidal zones. Thus; the choice of mangrove species to be planted is based on tidal inundation and land elevation in nearby natural mangrove forests.

Conclusion

Bioshields, no doubt play an imperative role in mitigating the effect of tsunamis and storm surges, but we must also consider if they are:

- ✰ Effective against extreme events
- ✰ Ecologically suitable
- ✰ Done with the involvement of local community.

Mangrove plantations, in the end can effectively serve as bioshields against hurricanes, tsunamis and storm surges as there are a wide variety of mangrove species available for cultivation depending on the nature of the tidal zone. Exotic plants such as *Casuarina equisetfolia L.,Tamarix Galicia L.,*and *Pinas* spp, produce only a slight reduction in damage because they are effective only against tidal waves of low height. Not only that, the other ecosystems, such as sand dunes(which also act as "shelterbelts") are destroyed during their cultivation.

References

Feagin, R.A., Mukherjee, N., Shanker, K. *et al.* (2010), "Shelter from the storm? Use and misuse of coastal vegetation bioshields for managing natural disasters", Volume-1, Page 2.

V. SelvamT. RavishankarV.M. Karunagaran R.Ramasubramanian,P. Eganathan,A. K. Parida (2005), "Toolkit for establishing Bioshields", M.S Swaminathan Research Foundation, Volume-05, Pages 7, 8, 14,15,16

R.S Bhalla (2007), "Do bio-shields affect tsunami inundation?", Foundation for Ecological Research, Advocacy and Learning, Volume-93, Page 831

Durganand Balsavar (2011), "Coastlines and Bio-Shields", The Hindu Survey of the Environment 2011, Pages 153, 154.

F. Dahdouh Guebas, L.P.Jayatissa, D.Di Nitto, J.O. Bossire, D. Lo Seen, and L.O. Koedam (2005), "How effective were mangroves as a defence against the recent tsunami?",Volume-15, Page 12

2013, Impact of Global Climate Change on Earth Ecosystems *Pages 155–171*
Editors: **D.R. Khanna, A.K. Chopra, Gagan Matta, Vikas Singh & Rakesh Bhutiani**
Published by: **BIOTECH BOOKS, NEW DELHI**

Chapter 21

Hydrochemistry and Generic Diversity of Aquatic Insects of Kurangani and Valiparai Streams of Western Ghats, Theni District, Tamil Nadu, India

T. Kubendran, C. Balasubramanian
T. Rathinakumar and P. Vinayaga Moorthi
Department of Zoology,
Thiagarajar College, Madurai – 625 009

Temperature is the key facto to predict the global climate change. The present study exploring the effect of increasing temperature on the fresh water aquatic insects communities, the indicator of water quality and physical disturbances by human interference. The physico-chemical parameters exemplify that, Kurangani stream had higher dissolved oxygen and optimum water and atmospheric temperature when compared to Valiparai stream. The aquatic insect composition unveiled that a total of 2001±0.81 and 1926±0.81 individuals belonging to 8 orders, 17 families and 21 genera were collected during March 2010 to February 2011 with respect to Kurangani and Valiparai stream. Functional feeding groups of the collected aquatic insect showed the dominance of Collectors followed by Gatherers

and Filterers in the above selected streams. BMWP- ASPT Score value portrays the health and portability of the Kurangani and Valiparai streams were good. Diversity indices were analyzed with help of PAST software. This study would enunciate the susceptibility of the aquatic insects of Valiparai stream, towards global climate change.

Keywords: *Aquatic macroinvertebrates, BMWP-ASPT, PAST and Diversity indices.*

Introduction

Natural water bodies like lakes, rivers, streams and groundwater should contain water of good quality because they are the only natural water sources on which life depends (Mann and Williamson, 1986). The industrialization and the rapid increase in human population have resulted in transformation of the natural environment. The environment has become hostile, posing many threats to health and welfare because of pollutants released into the environment. It should therefore be made safe and turned to good use for better standard of living and wealth creation (Thomas, 1972).

Aquatic insects are playing an important role in the cycle of materials and in trophic transfers (Cummins, 1974; Cummins *et al.*, 1989). Biological monitoring methods using aquatic insects have been developed and reliably tested in both temperate and tropical aquatic systems (Resh, 1979; Sivaramakrishnan *et al.*, 1996). Comprehensive information on the physico-chemical and biological parameters will aid in developing conservation strategies for the stream and riverine ecosystem of trophical biodiversity hot spots such as Western Ghats. Benthic macroinvertebrates are one of the most common groups of organisms used to assess the health of aquatic ecosystem (Rosenberg and Resh, 1993). Hence the present study carried out to evaluate the water quality using aquatic insects as bioindicator tools in the selected stream of Kurangani (Western Ghats) Tamil Nadu, India.

Materials and Methods

The Western Ghats are located between latitude 9′ 39′ and 10′ 30′ North and longitude 77′ 00′ and 78′ 30′ of East. For this investigation, streams of two hills namely Kurangani (11°N and 70° 50°E) was an amazing hill spot which was situated in Theni district, Thamilnadu, India. It is about 16 km away from Bodinayakkanur. Kurangani region is a sea of ridges and mountain peaks separated by valleys filled with virgin forests of silver oaks and eucalyptus and various plantations like tea, coffee and cardamom. It is about an attitude of 6,500 feet on the southern flank of the Western Ghats and Valipparai (11°N and 70° 50°E) was situated in Theni district, Tamilnadu, India. It is about 45 km away from Theni. It is about 6000 feet on the southern flank of the Western Ghats were chosen the investigation. These two streams are more important water source for Vaigai reservoir.

The study was conducted between August 2009 and July 2010. Hydrographic data were recorded by analysis of water samples following standard methods. Water and atmospheric temperatures was measured using accurate centigrade thermometer. Water current was measured by using cark. Stream width and depth were measured by using measuring tape. The hydrogen ion concentration (pH) of the water was estimated using 'Elico' pH meter. Dissolved oxygen in the water samples were estimated by the modified Winkler method. Free CO_2, Total solids, dissolved solids and suspended solids were assessed following methods prescribed by American Public Health Association (APHA) 1995.

Aquatic insects were sampled by taking kick net samples (mesh size 0.5-1.0cm) for minutes adopting the procedure described by Balasubramanian *et al.* (1992). Classification of aquatic insects into functional feeding groups follows Sivaramakrishnan (1989). Information is supplemented with relevant data from ready-made table given is Merritt and Cummins (1984). Tropic categorization was based on the general category of functional feeding groups of aquatic insects. Biological Monitor Working Party (BMWP) score may otherwise be known as TSS (Total Site Score). Families of aquatic insects were tabulated separately each station. Each family was ascribed the suitable score (10-1) depending on its sensitivity to pollution as prescribed by BMWP score system. Scores of indicator families were just added to arrive at BMWP or TSS (Armitage *et al.*, 1983). Derivation of Average Score Per Taxon (ASPT): ASPT value was obtained from TSS by dividing it with total number of families recorded as follows, ASPT= Total Site Score/Total number of families.

Alpha diversity indices of Shannon-Weiner index and Simpson index were worked out. The Shannon index and Simpson index of diversity were calculated (Ludwig and Reynolds, 1988). Similarities in taxonomic composition were quantified using Haccard's index (Sneath and Sokal, 1973) based on a presence-absence matrix for the insect fauna of each stream. A Bray- Curtis cluster analysis was performed using a flexible method (ß= -0.25) using PAST Programe (version 1.42). BMWP (Biological Monitor Working Party) score for each family, and then add the scores together to arrive at a site score. Score value for individual family reflects its pollution tolerance based on current knowledge of distribution and abundance (Armitage *et al.*, 1983).

Results and Discussion

The pH of Kurangani and Valiparai streams are alkalinic and ranges between 7.4-7.90; 6.85-7.70 respectively. But Sivaramakrishnan *et al.* (1990) and Joshi and Tyagi 1997 have reported different trend in a stream of cardamom hills, Western Ghats and Chirapani stream in Kumaon Himalayas. Dissolved oxygen (DO) concentration ranges between 7.64-9.7mg/L (Kurangani stream) 6.6-8.2mg/L (Valiparai stream) wide variation in DO concentration was reported for several South Indian streams in relation to season. Stream orders etc., Sivaramakrishnan *et al.* (1995) have studied DO concentration in tributaries of Kaveri river with different stream orders indicated a range between 4-8mg/L of DO. Investigation of free CO_2 on both Kurangani and Valiparai streams were maximum 1.3, 0.9mg/L and 0.7-1.1mg/L respectively. Total solids ranges from 0.03-0.75 and 0.2-2.8mg/L respectively. Dissolved solids range from 0.04-0.1mg/L and 0.02-0.2mg/L respectively.

Table 21.1(a): Order, Family and Genera of the Aquatic Insect Collected at Kurangani Stream, Southern Western Ghats during August 2009 to January 2010

Order	*Family*	*Genera*	*Aug*	*Sep*	*Oct*	*Nov*	*Dec*	*Jan*
Ephemeroptera	Beatidae	*Beatis* sp.	33±0.81	46±0.81	18±0.5	53±0.81	28±0.81	40±0.81
	Leptophlebiidae	*Choroterpes* sp.	24±0.47	23±0.47	2±0.47	16±0.47	3±0.47	2±0.47
	Caenidae	*Caenis* sp.	7±0.47	4±0.47	5±0.47	29±0.47	10±0.47	30±0.81
	Heptageniidae	*Epeorus* sp.	0±00	0±00	1±0.5	0±00	0±00	2±0.47
	Tricorythidae	*Tricorythodes* sp.	8±0.47	14±0.99	11±0.47	10±0.47	7±0.47	15±0.99
	Ephemeridae	*Ephemera* sp.	20±0.47	10±0.47	10±0.47	9±0.47	4±0.47	4±0.47
Plecoptera	Perlidae	*Neoperla* sp.	8±0.47	32±0.47	1±0.5	26±0.81	2±0.47	18±0.47
Trichoptera	Hydropsychidae	*Homoplectra* sp.	1±0.47	0±00	0±00	5±0.47	0±00	1±0.94
		Hydropsyche sp.	20±0.47	25±0.94	24±0.81	38±0.47	19±0.47	32±0.94
		Potamiya sp.	12±0.47	8±0.47	0±00	7±0.47	9±0.47	1±0.47
		Macronema sp.	6±0.47	12±0.47	0±00	11±0.47	0±00	1±0.47
		Leptonema sp.	0±00	0±00	0±00	0±00	0±00	1±0.94
	Polycentropodidae	*Polycentropus* sp.	2±0.47	3±0.47	4±0.47	3±0.47	1±0.47	1±0.47
	Stenopsychidae	*Stenopsyche* sp.	20±0.47	10±0.47	2±0.5	10±0.47	5±0.47	2±0.47
Diptera	Simulidae	*Simulium* sp.	23±0.47	4±0.47	0±00	2±0.47	5±0.47	3±0.47
	Chironomidae	*Chironomous* sp.	1±0.47	1±0.81	6±1.00	0±00	4±0.94	4±0.47
Coleoptera	Hydrophilidae	*Hydrobiomorpha* sp.	30±0.47	7±0.47	4±0.47	1±0.47	10±0.47	5±0.47
	Psephenidae	*Ectopria* sp.	1±0.47	0±00	0±00	1±0.47	0±00	2±0.47
Hemiptera	Gerridae	*Gerris* sp.	22±0.47	30±0.94	33±0.81	30±	25±0.81	28±0.81
Odonata	Coenagryonidae	*Coenagryon* sp.	2±0.47	3±0.47	2±0.47	2±0.47	2±0.47	4±0.47
Lepidoptera	Pyralidae	*Petrophilla* sp.	2±0.47	1±0.47	0±00	0±00	1±0.47	0±00

Table 21.1(b): Order, Family and Genera of the Aquatic Insect Collected at Kurangani Stream, Southern Western Ghats during February 2010 to July 2010

Order	*Family*	*Genera*	*Feb*	*Mar*	*Apr*	*May*	*Jun*	*July*
Ephemeroptera	Beatidae	*Beatis* sp.	35±0.81	30±0.81	40±0.81	30±0.81	25±0.81	20±0.81
	Leptophlebiidae	*Choroterpes* sp.	4±0.47	2±0.47	1±0.47	10±0.81	7±0.81	3±0.81
	Caenidae	*Caenis* sp.	28±0.81	10±0.47	25±0.81	18±0.47	10±0.81	5±0.81
	Heptageniidae	*Epeorus* sp.	2±0.47	4±0.47	5±0.47	0±0	18±0.81	4±0.81
	Tricorythidae	*Tricorythodes* sp.	3±0.47	1±0.47	4±0.47	2±0.47	2±0.47	3±0.47
	Ephemeridae	*Ephemera* sp.	0±0	1±0.47	1±0.47	4±0.47	7±0.81	2±0.81
Plecoptera	Perlidae	*Neoperla* sp.	10±0.81	5±0.81	3±0.81	10±0.81	25±0.81	10±0.81
Trichoptera	Hydropsychidae	*Homoplectra* sp.	1±0.47	0±0	1±0.47	1±0.47	1±0.47	0±0
		Hydropsyche sp.	30±0.81	15±0.81	10±0.81	30±0.47	18±0.81	26±0.81
		Potamiya sp.	1±0.47	0±0	1±0.47	0±0	1±0.47	2±0.47
		Macronema sp.	0±0	0±0	0±0	1±0.47	1±0.47	1±0.47
		Leptonema sp.	1±0.47	0±0	1±0.47	0±0	1±0.47	1±0.47
	Polycentropodidae	*Polycentropus* sp.	2±0.47	1±0.47	0±0	1±0.47	1±0.47	1±0.47
	Stenopsychidae	*Stenopsyche* sp.	8±0.81	4±0.81	15±0.81	3±0.47	8±0.81	10±0.81
Diptera	Simulidae	*Simulium* sp.	2±0.47	1±0.47	1±0.47	15±0.81	1±0.47	17±0.47
	Chironomidae	*Chironomous* sp.	0±0	0±0	0±0	0±0	0±0	0±0
Coleoptera	Hydrophilidae	*Hydrobiomorpha* sp.	3±0.81	2±0.81	10±0.81	15±0.81	1±0.47	3±0.47
	Psephenidae	*Ectopria* sp.	4±0.47	3±0.47	7±0.47	2±0.47	1±0.47	3±0.47
Hemiptera	Gerridae	*Gerris* sp.	15±0.81	21±0.81	15±0.81	11±0.81	10±0.81	25±0.81
Odonata	Coenagryonidae	*Coenagryon* sp.	2±0.47	1±0.47	3±0.47	3±0.47	1±0.47	0±0
Lepidoptera	Pyralidae	*Petrophilla* sp.	0±0	1±0.47	0±0	0±0	1±0.47	1±0.47

Table 21.1(c): Physico-chemical Parameters of the Selected Steam of Kurangani of Western Ghats during February 2010 to July 2010

Parameters	*Aug*	*Sep*	*Oct*	*Nov*	*Dec*	*Jan*	*Feb*	*Mar*	*Apr*	*May*	*June*	*July*
Water Temperature (°C)	21±0.47	23±0.47	23±0.47	22±0.94	21±0.47	22±0.47	22±0.94	21±0.47	22±0.47	23±0.94	22±0.47	21±0.47
Air Temperature (°C)	26±0.94	26±0.47	27±0.47	25±0.94	24±0.47	24±0.47	24±0.47	25±0.47	26±0.94	26±0.47	25±0.94	25±0.94
Water Current (M/Sec)	7.0±0.47	7.4±0.12	7.85±0.47	8.00±0.47	7.65±0.47	8.2±0.47	6.83±0.94	7.2±0.47	6.5±0.21	7.8±0.47	7.0±0.47	6.8±0.47
Width (M)	8.7±0.12	7.4±0.12	6.9±0.47	7.5±0.47	7.8±0.47	7.5±0.12	7.9±0.94	8.0±0.47	7.5±0.21	7.5±0.21	7.9±0.21	8.2±0.21
Depth (M)	0.3±0.12	0.5±0.08	0.6±0.12	0.8±0.47	0.8±0.47	0.8±0.47	0.6±0.12	0.8±0.47	0.5±0.21	0.6±0.21	0.8±0.21	0.8±0.21
Dissolved oxygen (mg/L)	7.64±0.27	9.07±0.03	7.79±0.12	7.28±0.12	8.5±0.47	7.9±0.12	8.2±0.12	8.0±0.47	8.2±0.47	8.0±0.47	9.7±0.47	9.0±0.47
Free CO_2	1.1±0.21	1±0.04	1±0.47	1±0.47	1±0.47	1±0.21	1±0.21	1±0.21	1±0.07	1.3±0.07	0.9±0.07	1±0.21
pH	7.63±0.27	7.93±0.06	7.8±0.12	7.85±0.12	7.8±0.470.12	7.5±0.12	7.8±0.47	7.2±0.21	7.4±0.47	7.5±0.47	7.1±0.21	7.7±0.47
Dissolved Solids	0.04±0.07	0.1±0.04	0.05±0.07	0.09±0.07	0.04±0.07	0.02±0.07	0.02±0.07	0.01±0.07	0.02±0.07	0.02±0.07	0.01±0.07	0.04±0.07
Suspended Solids	0.36±0.07	0.46±0.04	0.45±0.21	0.06±0.07	0.56±0.07	0.18±0.07	0.06±0.07	0.08±0.07	0.06±0.07	0.04±0.07	0.05±0.07	0.05±0.04

Table 21.1(d): Functional Feeding Groups Arrangement of Aquatic Insects Collected in Kurangani Stream, Southern Western Ghats during August 2009–July 2010

Order	*Family*	*Genera*	*Aug*	*Sep*	*Oct*	*Nov*	*Dec*	*Jan*	*Feb*	*Mar*	*Apr*	*May*	*Jun*	*July*
Ephemeroptera	Beatidae	*Beatis* sp	Coll, Ga, Scr	Coll, Ga, Scr	Coll, Ga, Scr	Coll, Ga, Scr	Coll, Ga, Scr	Coll, Ga, Scr	Coll, Ga, Scr	Coll, Ga, Scr	Coll, Ga, Scr	Coll, Ga, Scr	Coll, Ga, Scr	Coll, Ga, Scr
	Leptophlebiidae	*Choroterpes* sp	Coll, Ga, Scr	Coll, Ga, Scr	Coll, Ga, Scr	Coll, Ga, Scr	Coll, Ga, Scr	Coll, Ga, Scr	Coll, Ga, Scr	Coll, Ga, Scr	Coll, Ga, Scr	Coll, Ga, Scr	Coll, Ga, Scr	Coll, Ga, Scr
	Caenidae	*Caenis* sp	Coll, Ga	Coll, Ga	Coll, Ga	Coll, Ga	Coll, Ga	Coll, Ga	Coll, Ga	Coll, Ga	Coll, Ga	Coll, Ga	Coll, Ga	Coll, Ga
	Heptageniidae	*Epeorus* sp			Coll, Ga, SCR			Coll, Ga, SCR			Coll, Ga, SCR			Coll, Ga, SCR
	Tricorythidae	*Tricorythodes* sp												
	Ephemeridae	*Ephemera* sp												
Plecoptera	Perlidae	*Neoperla* sp	Pre, Eng	Pre, Eng	Pre, Eng	Pre, Eng	Pre, Eng	Pre, Eng	Pre, Eng	Pre, Eng	Pre, Eng	Pre, Eng	Pre, Eng	Pre, Eng
Trichoptera	Hydropsychidae	*Homoplectra* sp	Coll, Filtr			Coll, Filtr		Coll, Filtr	Coll, Filtr			Coll, Filtr		Coll, Filtr
		Hydropsyche sp	Coll, Filtr	Coll, Filtr	Coll, Filtr	Coll, Filtr	Coll, Filtr	Coll, Filtr	Coll, Filtr	Coll, Filtr	Coll, Filtr	Coll, Filtr	Coll, Filtr	Coll, Filtr
		Potamiya sp	Coll, Filtr	Coll, Filtr		Coll, Filtr	Coll, Filtr	Coll, Filtr	Coll, Filtr	Coll, Filtr		Coll, Filtr	Coll, Filtr	Coll, Filtr
		Macronema sp	Coll, Filtr	Coll, Filtr		Coll, Filtr		Coll, Filtr	Coll, Filtr	Coll, Filtr		Coll, Filtr		Coll, Filtr
		Leptonema sp						Coll, Filtr						Coll, Filtr

Contd...

Table 21.1(d)–Contd...

Order	*Family*	*Genera*	*Aug*	*Sep*	*Oct*	*Nov*	*Dec*	*Jan*	*Feb*	*Mar*	*Apr*	*May*	*Jun*	*July*
	Polycentropodidae	*Polycentropus* sp												
	Stenopsychidae	*Stenopsyche* sp	Coll	Coll	Coll	Coll	Coll	Coll	Coll	Coll	Coll	Coll	Coll	Coll
Diptera	Simulidae	*Simulium* sp	Coll, Filtr	Coll, Filtr		Coll, Filtr	Coll, Filtr	Coll, Filtr	Coll, Filtr	Coll, Filtr		Coll, Filtr	Coll, Filtr	Coll, Filtr
	Chironomidae	*Chironomous* sp	Coll, Ga	Coll, Ga	Coll, Ga		Coll, Ga	Coll, Ga	Coll, Ga	Coll, Ga	Coll, Ga		Coll, Ga	Coll, Ga
Coleoptera	Hydrophilidae	*Hydrobiomorpha* sp												
	Psephenidae	*Ectopria* sp												
Hemiptera	Gerridae	*Gerris* sp												
Odonata	Coenagryonidae	*Coenagryon* sp												
Lepidoptera	Pyralidae	*Petrophilla* sp												

Table 21.2(a) Order, Family and Genera of the Aquatic Insect Collected at Valiparai Stream, Southern Western Ghats during August 2009–January 2010

Order	*Family*	*Genera*	*Aug*	*Sep*	*Oct*	*Nov*	*Dec*	*Jan*
Ephemeroptera	Beatidae	*Beatis* sp.	28±0.81	36±0.94	28±0.5	50±0.94	18±0.94	30±0.81
	Leptophlebiidae	*Choroterpes* sp.	24±0.47	23±0.47	0±00	19±0.47	13±0.47	22±0.47
	Caenidae	*Caenis* sp.	7±0.47	13±0.81	10±0.81	29±0.47	17±0.47	25±0.81
	Heptageniidae	*Epeorus* sp.	0±00	0±00	1±0.04	0±00	0±00	2±0.47
	Tricorythidae	*Tricorythodes* sp.	10±0.94	10±0.99	15±0.47	8±0.47	5±0.47	15±0.99
Plecoptera	Perlidae	*Neoperla* sp.	6±0.47	12±0.47	10±0.47	11±0.47	7±0.47	9±0.47
Trichoptera	Hydropsychidae	*Homoplectra* sp.	1±0.47	0±0.47	1±0.47	1±0.47	0±0.47	1±0.47
		Hydropsyche sp.	25±1.24	15±0.81	24±0.81	27±0.81	30±0.94	22±0.94
		Potamiya sp.	1±0.47	1±0.47	1±0.47	0±00	0±00	1±0.5
		Macronema sp.	1±0.47	1±0.47	1±0.47	1±0.47	0±00	0±00
		Leptonema sp.	1±0.47	1±0.47	1±0.47	0±00	0±00	1±0.5
	Polycentropodidae	*Polycentropus* sp.	0±00	1±0.47	0±00	0±00	0±00	0±00
	Stenopsychidae	*Stenopsyche* sp.	4±0.47	9±0.81	8±0.81	14±0.99	11±0.47	10±0.47
Diptera	Simulidae	*Simulium* sp.	1±0.47	2±0.81	9±0.81	0±00	0±00	0±00
	Chironomidae	*Chironomous* sp.	0±00	0±00	0±00	1±0.47	0±00	0±00
Coleoptera	Hydrophilidae	*Hydrobiomorpha* sp.	10±0.81	18±0.81	24±0.81	27±0.81	30±0.94	22±0.94
	Psephenidae	*Ectopria* sp.	1±0.81	2±0.47	5±0.81	0±00	1±0.81	2±0.47
Hemiptera	Gerridae	*Gerris* sp.	12±0.81	5±0.81	12±0.81	11±0.47	10±0.47	7±0.47
Odonata	Coenagryonidae	*Coenagryon* sp.	1±0.47	0±0.47	1±0.47	0±00	0±00	1±0.47
Lepidoptera	Pyralidae	*Petrophilla* sp.	1±0.47	00±0.47	00±0.47	0±00	0±00	1±0.47

Table 21.2(b) Order, Family and Genera of the Aquatic Insect Collected at Valiparai Stream, Southern Western Ghats during February 2010–July 2010

Order	*Family*	*Genera*	*Feb*	*Mar*	*Apr*	*May*	*Jun*	*July*
Ephemeroptera	Beatidae	*Beatis* sp.	25±0.81	11±0.94	13±0.81	8±0.94	15±0.81	17±0.94
	Leptophlebiidae	*Choroterpes* sp.	11±0.81	15±0.81	17±0.81	18±0.47	10±0.94	8±0.81
	Caenidae	*Caenis* sp.	14±0.94	10±0.81	15±0.81	22±0.81	14 ±0.47	22±0.94
	Heptageniidae	*Epeorus* sp.	2±0.47	1±0.47	1±0.47	3±0.47	1±0.47	1±0.47
	Tricorythidae	*Tricorythodes* sp.	11±0.94	17±0.94	10±0.81	15±0.94	13±0.81	12±0.94
Plecoptera	Perlidae	*Neoperla* sp.	2±0.47	4±0.47	2±0.47	5±0.94	1±0.47	1±0.47
Trichoptera	Hydropsychidae	*Homoplectra* sp.	0±00	1±0.47	0±00	1±0.47	1±0.47	1±0.81
		Hydropsyche sp.	20±0.81	27±0.94	36±0.81	22±0.47	29±0.81	22±0.94
		Potamiya sp.	1±0.47	2±0.47	1±0.47	7±0.47	3±0.47	1±0.47
		Macronema sp.	1±0.47	5±0.47	7±0.81	4±0.47	14±0.81	3±0.47
		Leptonema sp.	3±0.47	0±00	1±0.47	3±0.47	1±0.47	3±0.47
	Polycentropodidae	*Polycentropus* sp.	0±00	1±0.47	0±00	1±0.47	0±00	1±0.47
	Stenopsychidae	*Stenopsyche* sp.	5±0.47	8±0.94	6±0.47	5±0.47	3±0.47	5±0.47
Diptera	Simulidae	*Simulium* sp.	4±0.47	2±0.47	3±0.47	10±0.94	2±0.94	8±0.81
	Chironomidae	*Chironomous* sp.	0±00	1±0.47	0±00	0±00	0±00	0±00
Coleoptera	Hydrophilidae	*Hydrobiomorpha* sp.	15±0.94	4±0.47	16±0.94	31±0.94	28±0.94	27±0.81
	Psephenidae	*Ectopria* sp.	3±0.47	6±0.94	2±0.47	1±0.47	9±0.81	6±0.47
Hemiptera	Gerridae	*Gerris* sp.	11±0.81	8±0.94	11±0.47	7±0.94	15±0.94	15±0.81
Odonata	Coenagryonidae	*Coenagryon* sp.	1±0.47	0±00	0±00	0±00	1±0.47	0±00
Lepidoptera	Pyralidae	*Petrophilla* sp.	0±00	1±0.47	0±00	1±0.47	0±00	0±00

Table 21.2(c): Physico-chemical Parameters of the Selected Steam of Valiparai of Western Ghats during August 2009–July 2010

Parameters	*Aug*	*Sep*	*Oct*	*Nov*	*Dec*	*Jan*	*Feb*	*Mar*	*Apr*	*May*	*June*	*July*
Water Temprature (°C)	26±0.81	25±0.94	22±0.81	23±0.47	22±0.81	22±1.24	22±1.24	24±0.81	24±0.94	24±0.94	23±1.24	23±0.94
Air Temperature (°C)	30± 1.24	28±0.94	24±0.94	25±0.47	23±1.24	23±0.81	23±0.94	25±1.24	26±0.81	26±1.24	25±0.81	27±1.24
Water Current (M/Sec)	3±0.47	3.8±0.14	4.9±0.16	3.2±0.16	3.9±	3±0.47	5.5±0.16	5.0±0.47	6.3±0.16	6.5±0.47	6.4±0.16	6.0±0.16
Width (M)	6.5± 0.08	6.5±0.12	7.5±0.20	6.4±0.16	6.8±0.20	7.5±0.20	6.9±0.20	7.5±0.12	7.0±0.12	7.9±0.20	7.3±0.12	6.0±0.20
Depth (M)	0.24± 0.02	0.2±0.04	0.3±0.12	0.4±0.08	0.2±0.04	0.3±0.08	0.5±0.16	0.6±0.16	0.7±0.02	0.6±0.01	0.4±0.02	0.7±0.16
Dissolved cxygen (mg/L)	7.85±0.02	6.64±0.09	7.07±0.16	8.04±0.01	8.26±0.02	6.83±0.16	.0±0.09	6.6±0.09	8.2±0.01	8.6±0.01	7.5±0.09	7.7±0.01
Free CO_2	1±0.21	1±0.09	1±0.35	1.1±0.12	1±0.21	1.1±0.35	1±0.21	1±0.35	1±0.35	0.9±0.21	0.7±0.21	1±0.35
pH	6.85±0.08	7.0±0.47	7.05±0.02	6.9±0.28	7±0.02	7.47±0.28	7.01±0.02	7.5±0.28	7.7±0.02	7.2±0.28	7.6±0.02	7.2±0.02
Dissolved Solids	0.28±0.09	0.05±0.04	0.09±0.01	0.05±0.01	0.08±0.01	0.02±0.01	0.02±0.01	0.06±0.01	0.07±0.01	0.10±0.01	0.19± 0.004	0.03± 0.004
Suspended Solids	2.52±0.08	0.5±0.04	0.41±0.02	0.55±0.05	0.4±0.02	0.18±0.04	0.50±0.04	0.45±0.04	0.33±0.04	0.4±0.04	0.34±0.02	0.23±0.04

Table 21.2(d): Functional Feeding Groups Arrangement of Aquatic Insects Collected in Valiparai Stream, Southern Western Ghats during August 2009–July 2010

Order	*Family*	*Genera*	*Aug*	*Sep*	*Oct*	*Nov*	*Dec*	*Jan*	*Feb*	*Mar*	*Apr*	*May*	*Jun*	*July*
Ephemeroptera	Beatidae	*Beatis* sp.	Coll, Ga, Scr	Coll, Ga, Scr	Coll, Ga, Scr	Coll, Ga, Scr	Coll, Ga, Scr	Coll, Ga, Scr	Coll, Ga, Scr	Coll, Ga, Scr	Coll, Ga, Scr	Coll, Ga, Scr	Coll, Ga, Scr	Coll, Ga, Scr
	Leptophlebiidae	*Choroterpes* sp.	Coll, Ga, Scr	Coll, Ga, Scr	Coll, Ga, Scr	Coll, Ga, Scr	Coll, Ga, Scr	Coll, Ga, Scr	Coll, Ga, Scr	Coll, Ga, Scr	Coll, Ga, Scr	Coll, Ga, Scr	Coll, Ga, Scr	Coll, Ga, Scr
Caenidae	*Caenis* sp.	Coll,Ga	Coll, Ga	Coll, Ga	Coll, Ga	Coll, Ga	Coll, Ga	Coll, Ga	Coll, Ga	Coll, Ga	Coll, Ga	Coll, Ga	Coll, Ga	Coll, Ga
	Heptageniidae	*Epeorus* sp.			Coll, Ga, Scr			Coll, Ga, Scr			Coll, Ga, Scr			Coll, Ga, Scr
	Tricorythidae	*Tricorythodes* sp.												
	Ephemeridae	*Ephemera* sp.												
Plecoptera	Perlidae	*Neoperla* sp.	Pre, Eng	Pre, Eng	Pre, Eng	Pre, Eng	Pre, Eng	Pre, Eng	Pre, Eng	Pre, Eng	Pre, Eng	Pre, Eng	Pre, Eng	Pre, Eng
Trichoptera	Hydropsychidae	*Homoplectra* sp.	Coll, Filtr			Coll, Filtr		Coll, Filtr	Coll, Filtr			Coll, Filtr		Coll, Filtr
		Hydropsyche sp.	Coll, Filtr	Coll, Filtr	Coll, Filtr	Coll, Filtr	Coll, Filtr	Coll, Filtr	Coll, Filtr	Coll, Filtr	Coll, Filtr	Coll, Filtr	Coll, Filtr	Coll, Filtr
		Potamiya sp.	Coll, Filtr	Coll, Filtr		Coll, Filtr	Coll, Filtr	Coll, Filtr	Coll, Filtr	Coll, Filtr		Coll, Filtr	Coll, Filtr	Coll, Filtr
		Macronema sp.	Coll, Filtr	Coll, Filtr		Coll, Filtr		Coll, Filtr	Coll, Filtr	Coll, Filtr		Coll, Filtr		Coll, Filtr
		Leptonema sp.						Coll, Filtr						Coll, Filtr
	Polycentropodidae	*Polycentropus* sp.												

Contd...

Table 21.2(d)–Contd...

Order	*Family*	*Genera*	*Aug*	*Sep*	*Oct*	*Nov*	*Dec*	*Jan*	*Feb*	*Mar*	*Apr*	*May*	*Jun*	*July*
	Stenopsychidae	*Stenopsyche* sp.	Coll	Coll	Coll	Coll	Coll	Coll	Coll	Coll	Coll	Coll	Coll	Coll
Diptera	Simulidae	*Simulium* sp.	Coll, Filtr	Coll, Filtr		Coll, Filtr	Coll, Filtr	Coll, Filtr	Coll, Filtr	Coll, Filtr		Coll, Filtr	Coll, Filtr	Coll, Filtr
	Chironomidae	*Chironomous* sp.	Coll, Ga	Coll, Ga	Coll, Ga		Coll, Ga	Coll, Ga	Coll, Ga	Coll, Ga	Coll, Ga		Coll, Ga	Coll, Ga
Coleoptera	Hydrophilidae	*Hydrobiomorpha* sp.												
	Psephenidae	*Ectopria* sp.												
Hemiptera	Gerridae	*Gerris* sp.												
Odonata	Coenagryonidae	*Coenagryon* sp.												
Lepidoptera	Pyralidae	*Petrophilla* sp.												

Studies on aquatic insects are very much limited when compared to other animal group of both lentic as well as lotic ecosystem (Gopal and Zutshi, 1998). Species diversity pattern in selected streams of Western Ghats have been well studied (Nagendran, 2004) Subramanian and Sivaramakrishan, 2005). In Kurangani stream aquatic insects belonging to 22 genera, 17 families and 8 orders were collected during investigation were given in Table 21.1. In Valiparai stream aquatic insects belonging to 20 genera, 16 families and 8 orders collected during investigation were given in Table 21.2. Indian Ephemeroptera includes 13 families and the present study also reviews the dominance of Ephemeroptera, but next to Tricoptera only. This order is represented by 4 families in Courtallum stream (Sivaramkrishnan and Job 1981), 4 families in cardamom hills (Sivaramakrishnan *et al.*, 1990) and 7 families in Perumal Patha Odai (Nagendran 2007). The selected stream having maximum 17 families in Kurangani and 16 families in Valiparai. The order Tricoptera in India is represented by 18 families (Ghosh, 1991). Burton and Sivaramakrishnan 1993 have reported 4 families from Silent valley. Sivaramakrishnan *et al.* (2000) have recorded 8 families from 17 locations in south Western Ghats in India.

Considerable information on the diversity of Hemiptera in Western Ghats is available for instance, 2 families from Valley (Burton and Sivaramakrishnan, 1993), 2 families in Alagar hills (Dinakaran and Krishnankutty, 1997) and 3 families in Southern Western Ghats (Sivaramakrishnan *et al.*, 2000). However, in Gerridae, only one family has been recorded in the present study and Nagendran (2007) has also reported similar finding in Perumal Patha Odai. The order Coleoptera is represented by 5 families in the selected streams. The data on

Coleoptera was very much limited as reported by Dinakaran and Krishnankutty, 1997. According to Sivaramakrishnan *et al.* (2000) only one family is present in Alagar and 3 families in some selected segments of Western Ghats. Perilidae (Plecoptera) and Pyralidae (Lepidoptera) are totally absent in Alagar hills (Dinakaran and Krishnan, 1997) and in Sirumalai hills streams Nagendran *et al.*, 2002. In the present Perlidae was found in Kurangani stream and all months of investigation except October, November, January, February, April and May. Whereas Valiparai stream having the Perlidae present in all months of investigation except August, October, November, February, April, June and July.

In the present study order Diptera was represented by 2 families. The orders Odonata and Coleoptera were represented by 1 and 2 genera respectively. Nagendran and Smija (2002) have reported 2 families of Odonata in the hill streams of Theni and more representation of Odonata was observed by Nagendran *et al*,(2002) in Sirumalai hill stream. India harbours 34 genera of Hemiptera belonging to 12 families (Srivastava, 1991). It is represented by 11 genera in selected streams. Ten genera in Courtallum (Sivaramakrishnan and Job, 1981), genera in cardamom hill (Sivaramakrishnan *et al.*, 1990), 12 genera from 17 locations in South Western Ghats (Sivaramakrishnan *et al.*, 2000) and 4 genera from the streams of Silent Valley (Burton and Sivaramakrishnan, 1993) have been represented. This group is dominated quantitatively in stream of Sirumalai hill (Nagendran *et al.*, 2002). Sivaramakrishnan *et al.* (2002) have listed 9 genera from 17 locations in South Western India. In the present study the presence of genera, Gerris and Sigara in the selected streams of Palani hills.

Neoperla, the only genera so for recorded from the stream of Western Ghats (Burton and Sivaramakrishnan, 1993 and Sivaramakrishnan *et al.*, 2000) and also same results were observed in the present study. Burton and Sivaramakrishnan (1993) and Sivaramakrishnan *et al.* (2000) have reported only one genera in Simulium in the order Diptera, whereas 6 genera were recorded from the stream of Palani hills. Petropbila, the only one genera in the order Lepidoptera has been recorded in both streams of Western Ghats.

Studies on trophic categorization of aquatic insects in South Indian stream were meager. The aquatic insects collected in the selected steams of Palani hills Western Ghats were grouped on the basis of the table given by Merritt and Cummins (1984) and Dudgeon (1984). This is because according to Dudgeon (1984) gut content analysis is cecessary only when published information on feeding habit of a given genera was inadequate. Percentage of trophic categories of aquatic insects in selected streams of Kurangani and Valiparai, Theni district, Tamil Nadu, India. In the present study, Collectors dominance at both sits among streams communities and their dominance increases with steam width. Sivaramakrishnan *et al.* (2000) have also reported the dominance of collectors at 17 locations of Southern Western Ghats.

The Biological Monitoring Working Party (BMWP) score system was designed to give a broad indication of the pollution of fivers throughout UK and Armitage *et al.*,(1993) have examined the performance of the BMWP score using samples in 41 rivers in England and Wales. In India, BMWP, ASPT score system was recommended as best for evaluating water quality in streams of Shillong (Gupta, 1994) and South Indian river Kaveri (Sivaramakrishnan *et al.*, 1996). An extensive study was conducted in 31 streams of Southern Western Ghats by Nagendran 2004; Nagendran *et al.*, 2002; Nagendran, 2007). The present study reveals that, the Total site score was maximum in Kurangani stream and minimum in Valiparai Stream. But the ASPT value was equal in both streams. Thus the selected streams were suffered due to organic pollution that reduce the abundant but not diversity.

Acknowledgements

Autours thanks to the Président Karumuthu. T. Kannan, Thiagrajar College for providing lab facilites and Authors P.V Moorthi and Thangavel Kubendran are thanks to UGC- Rajiv Gandhi National Fellowship, New Delhi for Financial support.

References

Armitage, P.D., Moss, D., Right, J.F and Furse, M.T (1983). The performance of a new biological water quality score system based on macroinvertebrates over a wide range of unpolluted running water sites. *Wat. Res.*, 17: 333-347.

Balasubramanian, C., Venkatraman, K. and Sivaramakrishnan,K.G. 1992. Biological studies on the burrowing mayfly. Ephemera nadinae McCafferty and Edmunds 1973 (Ephemeroptera: Ephemeridae) in Kurangani stream. J. Bombay Nat. His. Soc., 89: 72-77.

Burton, T.M. and Sivaramakrishnan, K.G. 1993. Comparition of the insect community in the streams of the Silent Valley national park in South India. Trop. Ecol., 34: 1-16.

Cummins, K. W. (1974). Structure and function of stream ecosystems. *BioScience, 24*: 631-644.

Cummins, K. W., Wilzbach, M. A., Gates, D. M., Perry, J. B. and Taliaferro, W. B. (1989). Shedders and riparian vegetation: leaf litter that falls into streams influences communities of stream invertebrates. *BioScience, 39*: 24-30.

Dinakaran, S and Kirushnankutty,K. 1997. A study on diversity of aquatic insects in a trophical stream of Alagar hill (Eastern Ghats). Proc. Seminar. On Biodiversity, K.V. College, Tanjore, India, 111-112.

Dudgeon, D. 1994. Research strategies for the conservation and management of tropical Asan stream and rivers. Int. J. Ecol. Environ. Sci., 20: 258-285.

Dudgeon, D. 1999. In tropical Asian streams- Zoorenthos, Ecology and consolation, Hong Kong University Press, Hong Kong, 828.

Dundgeon, D. 1984. Longitudinal and Temporal changes in functional organization of Macroinverebrates communities in the Lam Tsuen River, Hong Kong, Hydrobiologia, 111: 207-217.

Ghosh, D.K. 1991. Tricoptera. Anim. Resour. India, ZSI, Culcutta, 349-443.

Gopal, B and Zutshi, D.P. 1998. Fifty years of hydro biological research in India, Hydrobiologica, 384: 267-290.

Gupta, A. 1994. A Comparative assessment of methods for biological surveillance of running water quality using macroinvertebrates. Proc. Third. Nat. Symp. Environ. Department of Atomic Energy, Govt. of India,209-213.

Joshi, C.B and Tyagi, B.C. 1997. Some Limnological features of coldwater stream the Chiapani in Kumaon Himalaya, Uttarpradesh. J. Zool., 17(2): 156-162.

Mann, H.T. and Williamson, D. (1986). Water Treatment and Sanitation, Simple Methods for Rural Areas, pp. 6–7.

Merritt, R.W and Cummins, K.W. 1984. (Eds) An introduction to the Aquatic insects of North America, Kendal Hunt, Lowa, USA.

Merritt, R.W and Cummins, K.W. and Resh, V.H. 1984. Collective sampling and rearing methods for aquatic insects. In: An Introduction to the Aquatic insects of North American. (Ed) Merritt, R.W and Cummins, K.W. Kendall Hunt, Lowa, USA. 128-139.

Nagendran, N.A and Smika, M.S 2002. A seminar report on diversity of higher taxa of aquatic insects in three hill streams of Western Ghats a tropical forest stream of Sirumalai hill. South India and their use in Biomonitoring. In ecology of polluted waters (Ed) Kumar A.Ashish Publishing House, New Delhi, 655-657.

Nagendran, N.A. 2002. An Illustrated Ecology of Sirumalai Hill stream. A SERVE Pub., Madurai No. 1.10.

Nagendran, N.A. 2004. Biodiversity and Ecology of aquatic insect communities of hill streams of southern Western Ghats. Ph.D Thesis.Madurai Kamarajar University, 178.

Nagendran,N.A 2007. Stream characteristics and biomonitoring of Perumal Patha Odai, Karanthamaai, Tamilnadu, Ecol.Env&Cons., 13(1): 57-62.

Resh, V.H. (1979). Biomonitoring, species diversity indices and taxonomy. In: J.F. Grassle, G.P.Patil, W.K.Smith and C.Taillie, editors. *Ecological diversity in theory and practice.* 241-253. Fairland: *M.D.International Cooperative Publishing.*

Sivaramakrishnan, K.G and Job, S.V 1981. Studies on Mayfly populations of Courtallam stream. Proc. Symp.Ecol. Anim. Popul. ZSI. 105-116.

Sivaramakrishnan, K.G., Hannaford, M.J and Resh, V.H. 1996. Biological assessment of the Kaveri river catchment, South India using benthic macroinvertebrates: applicability of owater quality monitoring approaches developed in other countries. Int. J. Eco. Environ. Sci., 22: 113-132.

Sivaramakrishnan, K.G., Sridhar, S and Marimuthu,S. 1995. Spatial patterns of benthic macroinvertebrates distribution along river Kaveri and its tributaries (India). Int.J.Ecol.Environ. Sci., 21:141-166.

Sivaramakrishnan, K.G., Sridhar,S and Venkataraman, K. 1990. Habits, Micro distribution, Life cycle pattrns and Trophic relationships of Mayflies of Cardamom hills of Western Ghats. Hexapoda, 2: 9-16.

Sivaramakrishnan, K.G., Venkataraman, K., Moorthy, R.K., Subramanian,K.A and Utkarsh, G. 2000. Aquatic insect diversity and use in the streams of Western Ghats, India. Indian. Ins, Sci., 80: 978-987.

Sivaramakrishnan,K.G and Venkatramam, K 1990. Abundance, altitudinal distribution and swarming of Ephemeroptera in Palani hill, South India. In: Mayflies and Stone flies (Ed Campbell, I.C.)Kluwer, New York, Pp: 209-213.

Thomas, V.R. (1972). Systems Analysis and Water Quality Management. first ed., New York.

2013, Impact of Global Climate Change on Earth Ecosystems *Pages* ***173–178***
Editors: **D.R. Khanna, A.K. Chopra, Gagan Matta, Vikas Singh & Rakesh Bhutiani**
Published by: **BIOTECH BOOKS, NEW DELHI**

Chapter 22

Conservation of Threatened Biodiversity: A Today's Need!

V.A. Shende[1] and K.S. Janbandhu[2]

[1]*K.Z.S. Science College, Bramhani-Kalmeshwar, Dist. Nagpur – 441 501, M.S.*

[2]*Department of Zoology, Institute of Science, R.T. Marg, Nagpur, M.S.*

Global biodiversity and other ecosystem services are under significant threat from the combined effects of human-induced climate and land-use change. Biodiversity influences ecosystem services, that is, the benefits provided by ecosystems, which contribute to making human life both possible and worth living. Its degradation is threatening the fulfillment of basic needs and aspiration of humanity as a whole. Changing temperature and precipitation regimes will interact with existing drivers such as habitat loss to influence species distributions despite their protection within reserve boundaries. As a consequence, the total number of species on the planet is decreasing subsequently. It is important that scientists and the public work together to coordinate the many plans of action, so that a cohesive program of action can be undertaken for the international common good. Nations represent the sovereign authorities that implement and manage conservation efforts.

Clearly, future land-use and climate change impacts will need to be integrated more prominently in all components of conservation action over the coming decades.

Keywords: *Biodiversity, Conservation, Climate change, Ecosystem.*

Introduction

The concept of biodiversity has provoked considerable debate and misunderstanding among the general public, decision-makers, and even the scientific community. Biodiversity is the degree of variation of life forms within a given ecosystem, biome or an entire planet. It is a measure of the health of ecosystem and a function of climate. Biodiversity includes all organisms, species, and populations; the genetic variation among these; and all their complex assemblages of communities and ecosystems. It also refers to the interrelatedness of genes, species, and ecosystems and their interactions with the environment.

The biodiversity loss is a significant issue for scientists and policy-makers and the topic is finding its way into living rooms and classrooms. Species are becoming extinct at the fastest rate known in geological history and most of these extinctions have been tied to human activity. Preservation of global biodiversity is a priority in strategic conservation plans that are designed to engage public policy and concerns affecting local, regional and global scales of communities, ecosystems and cultures. Action plans identify ways of sustaining human well-being, employing natural capital, market capital and ecosystem services.

Environmental Services

Biodiversity helps to maintain the gaseous composition of the atmosphere and thus to regulate climate. Biodiversity influences ecosystem services, that is, the benefits provided by ecosystems to humans that contribute to making human life both possible and worth living. There are two categories of contributions providing to the animals *i.e.* material goods and environmental services. The material goods has been frequently and widely documented (Myers, 1983; McNeely, 1988; Oldfield, 1989; Ehrlich and Ehrlich, 1992), principally in the form of new and improved foods, medicines and drugs, raw materials for industry and sources of bio-energy. The environmental services has been far less documented even though the issue was identified as unusually significant almost two decades ago and even though its total value is surely far greater than that of the material goods (Bishop, 1987; Di Castri and Younis, 1990; Green and Tunstall 1991; Ehrlich and Ehrlich, 1992; Risser, 1995).

There are several categories of environmental services, including regulation of climate and biogeochemical cycles, hydrological functions, soil protection, crop pollination, pest control, recreation, research, ecotourism and a number of miscellaneous services (Ricklefs, 1987; Whelan, 1991; Schlesinger, 1991; Panayotou and Ashton, 1992; Pimentel *et al.*,, 1992, 1995; Eagles, Buse, and Hvengaard, 1993, Odum, 1993; Norman Myers, 1996). They can be defined as any functional attribute of natural ecosystems that are demonstrably beneficial to humankind (Cairns and Pratt, 1995).

The ecosystem services are comprise the main indirect values of biodiversity, as that of direct values in the form of material goods such as timber, fish, plant-based pharmaceuticals and germ-plasm infusions for major crops. Indirect values include generating and maintaining soils, converting solar energy into plant tissue, sustaining hydrological cycles, storing and cycling essential nutrients (nitrogen fixation), supplying clean air and water, absorbing and detoxifying pollutants, decomposing wastes, pollinating crops and other plants, controlling pests, running biogeochemical cycles (carbon, nitrogen, phosphorus, and sulfur), controlling the gaseous mixture of the atmosphere and regulating weather and climate at both macro and micro levels. Thus basically include three forms of processing, namely of minerals, energy, and water (Perrings, 1987).

There are clear implications of climate change for life on the mother earth. Many species depend on complex relationships between soils and several other species of animals and plants. Shifting of forest ranges would affect plants and animals. The future ranges of species will be dramatically different than today's (Smith and Tirpak, 1989).

Threat to Biodiversity and Humanity

Five main threats to biodiversity are commonly recognized in the programmes of work of the Convention: invasive alien species, climate change, nutrient loading and pollution, habitat change and overexploitation. Behind these direct drivers of biodiversity loss, there are a number of indirect drivers that interact in complex ways to cause human-induced changes in biodiversity. They include demographic, economic, socio-political, cultural, religious, scientific and technological factors, which influence human activities that directly impact on biodiversity.

We are clearly benefited from the diversity of organisms that to use for medicines, food, fibers and other renewable resources. In addition, biodiversity has always been an integral part of the human experience and there are many moral reasons to preserve it for its own sake. People who rely most directly on ecosystem services, such as subsistence farmers, the rural poor, and traditional societies, face the most serious and immediate risks from biodiversity loss (Jutro, 1991). Historically, direct threats to human health and the protection of our environment really were two different issues but they are directly related to each other.

The likelihood is that many species will disappear locally and many more will become extinct. Biodiversity on the planet, the diversity of habitat and ecosystems of species and genetic diversity will all be dramatically reduced. Such a warming would likely be accompanied by dramatic changes in global rainfall patterns (Mitchell *et al.*,, 1990).

Global climate change, brought about by rising levels of greenhouse gases, threaten ecosystems to alter the geographic distribution of many habitats and their component species (Burns *et al.*, 2003). Studies of paleobiology indicate that plants and animals are exquisitely sensitive to changes in climate (Delcourt, 1983; Davis, 1989). Recent empirical studies strongly suggest that wildlife species are already responding to recent global warming trends with significant shifts in range distribution and phenology (*e.g.*, earlier breeding, flowering, and migration). Several recent reviews and meta-analyses provide a synthesis of contemporary global warming effects on wildlife (Hughes, 2000; McCarthy, 2001; Walther *et al.*, 2002; Parmesan and Yohe, 2003; Root *et al.*, 2003) which can further disrupt current species associations.

The degradation of biodiversity is threatening the fulfillment of basic needs and aspiration of humanity as a whole, but especially and most immediately the most disadvantaged segments of society (Díaz *et al.*, 2006).

Coordinate the Action Plans

Fortunately, a number of biological scientists maintained a cross-disciplinary association with the climate scientists, especially a few with wide-ranging minds and as the climate science developed, so that they concern for the possible implications of climate change. Interdisciplinary scientific inquiry can help us identify and understand the nature of these issues and project what they may mean for life on earth under different scenarios, but science alone cannot provide adequate solutions.

For that, Social scientists have also worked with the natural scientists to develop those scenarios depending on assumptions about technological developments and the ability of different societies to make changes, some fundamental and some fairly simple. In arriving at solutions to these problems, we must look further afield at the linkages between environment and society and the value changes that may be necessary in order to fashion economic growth modes that do not ultimately jeopardize the health of the biosphere as a living system (Jutro, 1991). This interaction helped raise the public consciousness of the issue. Although the public was little taken with the esoteric aspects of atmospheric sciences, the concern with those implications was very real.

A very large and diverse public demonstrates a connection with nature and a sense of concern about environmental problems. The capacity of the public (and the media) to respond in a more massive and emphatic way to some environmental issues, such as global warming (Bowman, 2007), points the way for greater connections with the public on biodiversity issues. We must provide the enhanced understanding of biodiversity and its degradation in a way that empowers people to make choices and take action based on sound science and reliable recommendations. In the meantime, many avenues for attaining this goal-communications through media, environmental NGOs, contributions of public science institutions and the development of citizen science programs have been established (Novacek, 2008).

The resulting policies of conservation and sustainable development will depend on the struggle and negotiation over models of nature and social practice among the groups involved. The social movements considered here are those that explicitly construct a political strategy for the defense of territory, culture, and identity linked to particular places and territories. These movements enact a cultural politics that is mediated by ecological considerations.

Conclusion

The environmental services of biodiversity are certainly significant. Wild communities, for instance, are likely to dramatically change in composition as their individual components respond to climate change with different migration rates and changes in reproductive physiology. Worldwide biodiversity loss and climate change are issues of global scope and importance that have recently become subjects of considerable public concern. Both issues are intriguing, in that their perceived threat lies in their potential to disrupt ecological functioning and stability rather than from any direct threat they may pose to human health.

For the conservation of threatened biodiversity effective reserve administration and governance, better management of the surrounding matrix and improved regional/national/international coordination are required. Thus, it is, a unique time in science and society when we find this stunning convergence of opinion from many different directions, climatologists, biotechnologists, industrial entrepreneurs, and sustainable biosphere adherence proponents, calling for an international biodiversity initiative that encompasses exploration, inventorying, and conservation of the diversity of the planet.

Acknowledgement

Corresponding author wishes to acknowledge Dr. Anand Tikhe for valuable suggestions and guidance, Dr. Kishor Patil for his inspiring attitude and all well-wishers.

References

Bishop, R. C. (1987) Valuing Wildlife: Economic and Social Perspectives, eds. Decker, D. J. and Goff, D. R. (Westview, Boulder, CO), pp. 24-34.

Bowman K. (2007) Global warming: Not so hot. *Washington Post*. Available at www.washingtonpost.com/wp-yn/content/article/2007/08/30/AR2007083001440.html.

Cairns, J. and Pratt, J. R. (1995) Evaluating and Monitoring the Health of Large-Scale Ecosystems, eds. Rapport, C. L. and Calow, P. (Springer, Berlin), pp. 63-76.

Catherine E. Burns, Kevin M. Johnston, and Oswald J. Schmitz (2003) Global climate change and mammalian species diversity in U.S. national parks. *PNAS* vol. 100(20), 11474–11477

Davis, M. B. (1989) Insights from paleoecology on global change. *Ecol. Soc. Am. Bull.* 70 (4): 222-228.

Delcourt,H. R.,Delcourt,P. A.,andWebb, T.,Im. (1983) Dynamic plant ecology: the spectrum of vegetational change in space and time. *Q. Sci. Rev.*1:153-175.

Di Castri, F. and Younis, T. (1990) *Ecosystem Function of Biodiversity* (Int. Union Biol. Sci., Paris), *Biol. Int. Special Issue* 22.

Eagles, P. F. J., Buse, S. D. and Hvengaard, G. T., eds. (1993) The Ecotourism Society Annotated Bibliography (Ecotourism Soc.,Bennington, VT).

Ehrlich, P. R. and Ehrlich, A. H. (1992) *Ambio* 21, 219-226.

Green, C. H. and Tunstall, S. M. (1991) *J. Environ. Manage.* 33,123-141.

Hughes, L. (2000) *Trends Ecol. Evol.* 15, 56–61.

Jutro, P. R. (1991) Biological Diversity, Ecology, and Global Climate Change. *Environmental Health Perspectives*, 96, pp. 167-17d.

McCarthy, J. P. (2001) *Conserv. Biol.* 15, 320–331.

McNeely, J. A. (1988) Economics and Biological Conservation: Developing and Using Economic Incentives to Conserve Biological Resources (Int. Union for Conservation of Nature and Natural Resources, Gland, Switzerland).

M. J. Novacek* (2008) Engaging the public in biodiversity issues. *PNAS* vol. 105(1) 11571–11578.

Mitchell, J. F., Manabe, S., bkioda, T., and Meleshenko, V. (1990) Equilibrium climate change and its implications forthe future. In: Climate Change: The IPCC Scientific Assessnt. Cambridge University Press, Cambridge.

Myers, N. (1983) A Wealth of Wild Species (Westview, Boulder,CO).

Myers N. (1996) Environmental services of biodiversity, *Proc. Natl. Acad. Sci. USA*, Vol. 93, pp. 2764-2769.

Odum, E. P. (1993) Ecology and Our Endangered Life-Support Systems (Sinauer, Sunderland, MA).

Oldfield, M. L. (1989) The Value of Conserving Genetic Resources (Sinauer, Sunderland, MA).

Panayotou, T. and Ashton, P. S. (1992) Not By Timber Alone (Earthscan, London).

Parmesan, C. and Yohe, G. (2003) *Nature* 421, 37–42.

Perrings, C. (1987) Economy and Environment: A Theoretical Essay on the Interdependence of Economic and Environmental Systems (Cambridge Univ. Press, Cambridge, U.K.).

Pimentel, D., Harvey, C., Resosudarmo, P., Sinclair, K., Kurz, D., McNair, M., Crist, S., Shpritz, L., Fitton, L., Saffouri, R. and Blair, R. (1995) *Science* 267, 1117-1122.

Pimentel, D., Stachow, U., Takacs, D. A., Brubaker, H. W., Dumas, A. R., Meaney, J. J., O'Neil, J. A. S., Onsi, D. E. and Corzilius, D. B. (1992) *BioScience* 42, 354-362.

Ricklefs, R. E. (1987) *Science* 235, 167-171.

Risser, P. G. (1995) Conserv. Biol. 9, 742-746.

Root, T. L., Price, J. T., Hall, K. R., Schneider, S. H., Rosenzweig, C.andPounds, J. A. (2003) *Nature* 421, 57–60.

Díaz S., Joseph Fargione, F. Stuart Chapin III, David Tilman (2006) Biodiversity Loss Threatens Human Well-Being, *PLoS Biology*, Vol. 4 (8) 1300-1305.

Schlesinger, W. H. (1991) Biogeochemistry: An Analysis of Global Change (Academic, San Diego).

Smith, J. B., and Tirpak, D. (1989) Agriculture. In: The Poential Effects of Global Climate Change on the United States U. S. Environmental Protetion Agency. Washington, DC p. 155.

Walther, G.-R., Post, E., Convey, P. Menzel, A., Parmesan, C., Beebee, T. J., Fromentin, J.-M., Hoegh-Guldberg, O. and Berlein, F. (2002) *Nature* 416, 389–395.

Whelan, T., ed. (1991) Nature Tourism: Managing for the Environment (Island Press, Washington, DC).

2013, Impact of Global Climate Change on Earth Ecosystems *Pages* ***179–186***
Editors: **D.R. Khanna, A.K. Chopra, Gagan Matta, Vikas Singh & Rakesh Bhutiani**
Published by: **BIOTECH BOOKS, NEW DELHI**

Chapter 23

Population Dynamics of Zooplankton from the Two Lakes, Bothali (Mendha) and Murkhala, Dist. Gadchiroli, Maharashtra, India

Rajendra V. Tijare

Assistant Professor,
Govt. Institute of Science, Nagpur, Maharashtra

The spatial and temporal distribution of zooplankton in Bothali (Mendha) and Murkhala lakes were studied monthly from November 2003 to October 2005. Bothali (Mendha) lake located at 20°–11° latitude and 80′–04′ longitude besides Dhanora road, having an area about 39.68 hectors. Murkhala Lake located near Chandrapur road at 20°–39° latitude and 80′–96′ longitude.

Maximum population of zooplankton recorded in Murkhala lake than Bothali (Mendha) lake. Among the total zooplanktonic forms rotifers were counted highest which followed by crustaceans forms. Periodical appearance of nauplius larvae indicating the new recruitment in population.

From copepods, cyclopoides were dominant than calanoides. The cladocerans specially *Simocephalus vetulus*, *Alona rectangular* and *Bosminopsis deitersi* were found in only Bothali lake water while *Chydorus spharicus* found only in Murkhala lake water. Maximum species were recorded from family Brachionidae among the rotifers. The references on occurrence of zooplankton from freshwaters of this region are not available hence the study was undertaken.

***Keywords:** Population, Zooplankton, Bothali, Murkhala.*

Introduction

Bothali and Murkhala lake, located near about 05 and 07 Km. away from Gadchiroli city, situated in rural area. Both lakes are old "Malgujari" Talav type. Bothali lake having an area of about 39.68 hectares and Murkhala lake having 39.64 hectare area. Lake water is using mainly for irrigation and fishery purposes. Pollution in these lakes mainly due to anthropogenic activities, surface run-off and received untreated discharges including animal wastes. The lake water consists of phytoplankton (mainly members of Chlorophyceae and Bacillariophyceae), the aquatic vegetation of lake is basically submergrd *Hydrilla, Ceratophyllum, Utricularia* and emergent *Nymphea*.

Material and Methods

Plankton samples were collected by using conical plankton net with a bottle at the end. 200 meshes/cm. bolton silk conical plankton net used for sampling. 50-liter lake water was filtered through a net for quantitative analysis. For qualitative analysis, the net was towed horizontally, vertically, and obliquely. Counting of zooplankton was done by using Sedgwick Rafter cell method. Once the sample was collected, the organisms were preserved in 4 per cent formalin.

Observations

In the present investigations, an attempt has been made to study occurrence, distribution, and species number of zooplankton in two lakes. The qualitative and quantitative analysis of zooplankton communities was carried out at regular intervals during the years 2003-04 and 2004-05. The zooplanktonic species of Copepodes, Cladocerans, Rotifers, Ostracodes, with the nauplii larvae and eggs, which observed were recorded. Highest numbers of zooplanktonic species were recorded from Bothali lake in compare to Murkhala lake during both the study years respectively.

In all 34 species of zooplankton were recorded during the investigation period from both lakes. In Bothali lake rotifers were 40.61 per cent and 33 per cent which followed by copepods (19.97 per cent and 20.97 per cent) and cladocera (3.36 per cent and 9.32 per cent). The population of nauplius larvae was 34.10 per cent during first year, which declined to 33.3 per cent in second year of study. The egg constituted 01.63 and 3.18 per cent of the total population in both the respective study years, the eggs were observed only during May and July to October months of both the years.

Table 23.1: Observations

Sl.No.	Group	Name of Species	Bothali	Murkhala
A.	**Crustacea: Copepoda**	1. *Cyclops* sp.	+	+
		2. *Mesocyclops* sp.	+	+
		3. *Diaptomus* sp.	+	+
B.	**Crustacea: Cladocera**	1. *Cereodaphnia rigaudi* (Richard)	+	+
		2. *Moina macrura* (Jurine)	+	+
		3. *Simocephalus vetulus* (O.F.Muller)	+	–
		4. *Alona rectangula* (Sars)	+	–
		5. *Bosminopsis deitersi* (Richard)	+	–
		6. *Chydorus spharicus*	–	+
C.	**Rotifera**	1. Brachionus calyciflorus (Wierzeijski)	+	+
		2. *B. fulcatus* (Zacharias)	+	+
		3. *B. caudatus* (Borrois and Dadey)	+	+
		4. *B. angularis* (Gosse)	+	+
		5. *B. forficula* (Wierzeijski)	+	+
		6. *B. diversicornis* (Daday)	+	+
		7. *B. quadridentacus* (Hermann)	+	+
		8. *B. platulus platulus* (O.F.Muller)	+	+
		9. *Keratella tropica* (Apstein)	+	+
		10. *K. colchearis* (Gosse)	+	+
		11. *K. quadrata* (Muller)	+	–
		12. *K. ticinensis*	+	–
		13. *Asplanchna* sp.	+	+
		14. *Epiphanes macrourus* (Barois and Daday)	+	–
		15. *Filinia longiseta* (Ehrenberg)	+	+
		16. *Filinia apoloensis* (Zach.)	+	+
		17. *Polyarthra vulgaris* (Carlin)	+	+
		18. *Hexarthra* sp.	+	–
		19. *Tricocera cylindrical* (Imhoff)	–	+
		20. *Lecane bulla*	+	–
		21. *L. papuana*	+	–
		22. *L. eswari*	+	–
		23. *Ascomorpha* sp.	+	–
		24. *Anuareopsis fissa* (Gosse)	+	–
D.	**Ostracoda**	1. *Stenocypris hislopi*	+	+

+: Zooplanktonic fauna observed –: Zooplanktonic fauna not observed.

Large population of rotifers (52.51 per cent and 63.24 per cent) were counted from Murkhala lake water while Copepode population (16.68 per cent and 12.65 per cent) were recorded comparatively less than Bothali lake during both the investigation years. The population pattern of nauplius larvae was same that of Bothali lake.

The prominent rotifers found in these two lake waters were, *Brachionus* sp., *Keratella* sp., *Fillinia* sp., *Asplanchna* sp. and *Polyarthra* sp.

Result and discussion

The data collected for zooplankton are graphically represented in Figures 23.1–23.8. The major groups of zooplankton observed during the study, are Rotifera,

I. ZOOPLANKTON POPULATION IN BOTHALI LAKE

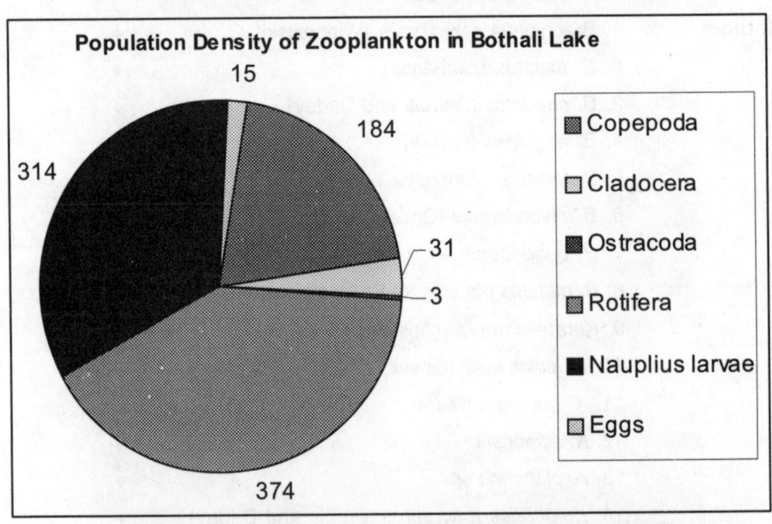

Figure 23.1: Species Population in Year 2003-04

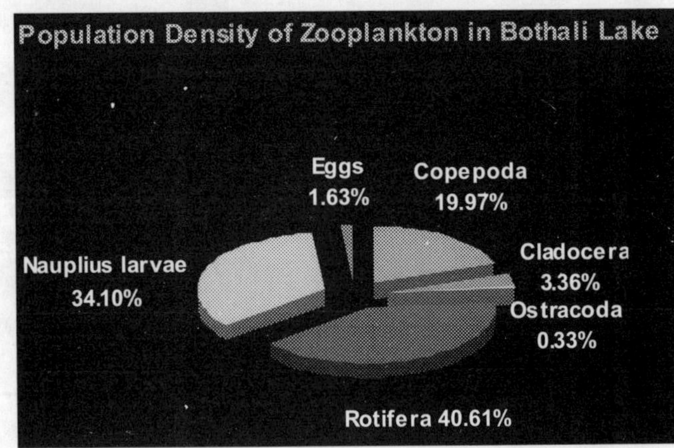

Figure 23.2: Percentage Population of Species 2003-04

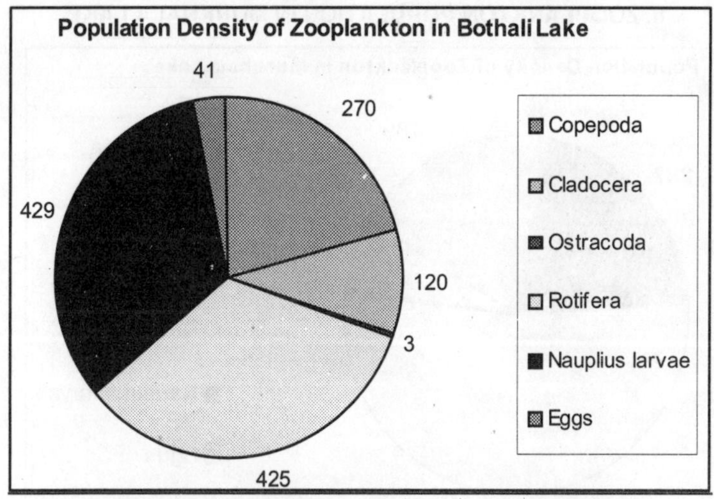

Figure 23.3: Species Population in Year 2004-05

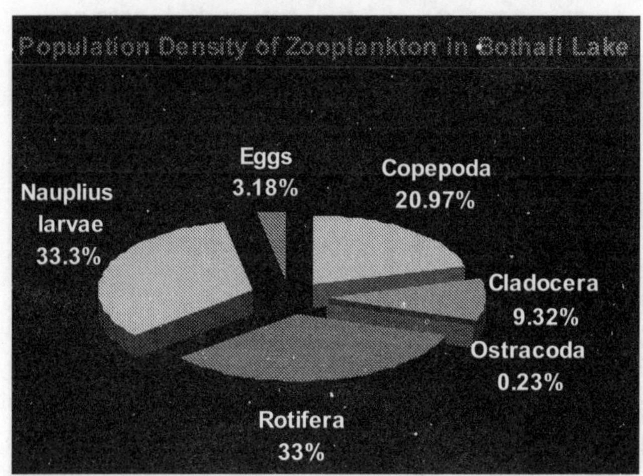

Figure 23.4: Percentage Population of Species 2004-05

Cladocera, Copepoda and Ostracoda, among all group Rotifera is dominant in the lake water, however there is no significant relationship between these groups of organisms.

In the present investigation, 24 species of Rotifera, 06 species of Cladocera, 03 species of Copepoda and 01 species of Ostracoda were recorded for the first time. Among rotifers *Brachionus* sp. and *Keratella* sp. are dominant and their presence is most of the time, throughout the year. The highest population of rotifers recorded in the summer and winter months (Somani *et al.*, 2003). The copepods represented by *Cyclops* sp. and *Diaptomus* sp. were observed throughout the study years in lake water. The second dominant group observed in both the lake water was of Nauplius larvae. Periodical appearance of nauplius larvae was seen, indicating the new

II. ZOOPLANKTON POPULATION IN MURKHALA LAKE

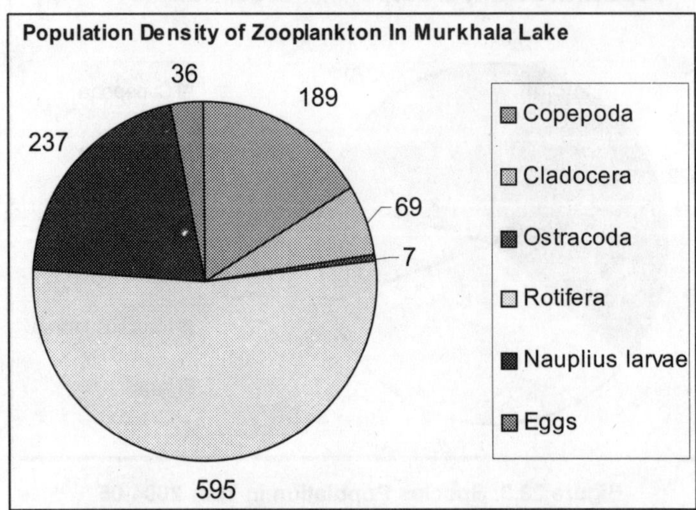

Figure 23.5: Species Population in Year 2003-04

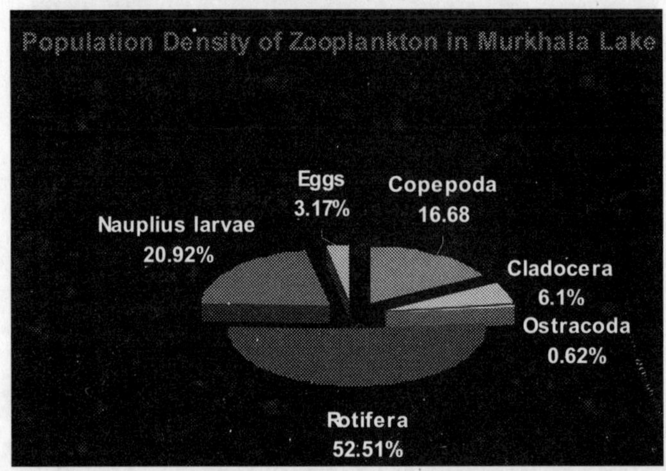

Figure 23.6: Percentage Population of Species 2003-04

recruitment in population (Patil, S.G., 1987). Population of copepods and cladocerans increases in second year especially in Bothali lake probably due to more favorable environmental conditions for reproduction of them.

Among cladocera *Cereodaphnia rigaudi* and *Moina macrura* found in both lake water. *Simocephalus vetulus, Alona rectangular,* and *Bosminopsis deitersi* recorded from Bothali lake while *Chydorus spharicus* observed in Murkhala lake only. The cladocerans did not show any fixed pattern of abundance during the study period.

Maximum species of rotifers recorded from Bothali lake. In Murkhala lake species like *Keratella quadrata, K. ticinensis, Epiphanes macrourus, Anuareopsis fissa, Hexarthra* sp., *Lecane* sp. and *Ascomorpha* sp. were not observed while *Tricocera cylindrical* was

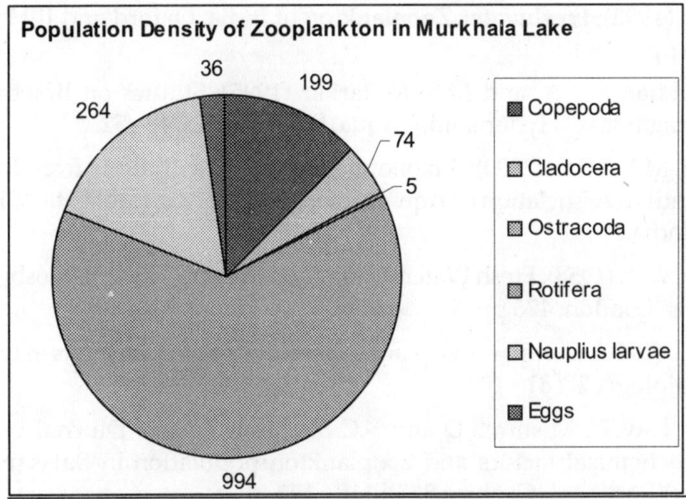

Figure 23.7: Species Population in Year 2004-05

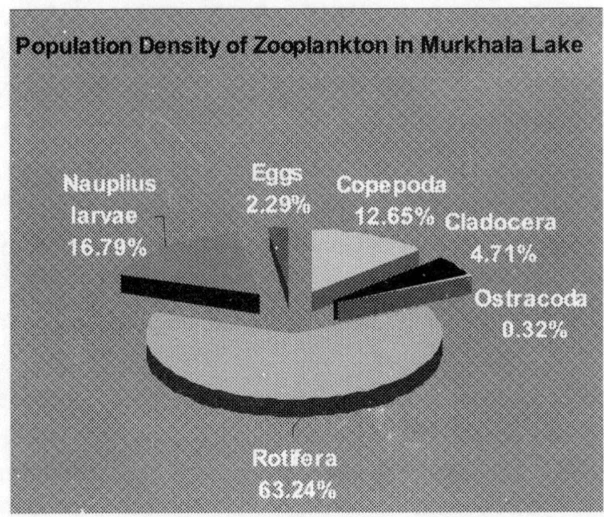

Figure 23.8: Percentage Population of Species 2004-05

recorded only from this waterbody. Only one ostracode *i.e. Stenocypris hislopi* found in both the lake water.

Acknowledgment

The authors are grateful to the Director, Institute of Science, Nagpur for providing necessary laboratory facilities and to scientist of ZSI, Western region, Pune, (Maharashtra) for helpful suggestions during identification of zooplankton.

References

Anderson, K.H. (1889): Notes on Indian Rotifers., J. Asiatic Soc. Bengal, Culcutta, 58: 345 - 358.

Battish, S.K. (1992): Freshwater Zooplankton of India.,Oxford and IBH Publishing Co.,223 pp.

Chandrashekhar, S.V.A. and M.S. Kodarkar (1995): Studies on Brachionus from Saroornagar lake, Hyderabad.J. Aqua. Bio. 10 (1&2): 48 - 52.

Dhanapathi, M.V.S.S.S. (2000): Taxonomic notes on The Rotifers from India. (1889-2000). Indian Association of Aquatic Biologist, (IAAB), Publication, Hyderabad (A.P.), India.

Edmondson, W.T. (1958): Fresh Water Biology (2nd edition), The C.V. Mosby Company, St. Louis, London: 426 pp.

George, M.G. (1966): Comparative plankton ecology of five fish tanks in Delhi, India., Hydrobiologia 27: 81 - 108.

Jakher,G.R., Day,T., Mishra,S.D and S.C.Bhargava (1981): Diurnal variations in physico-chemical factors and zooplankton population in Balasamand Lake, Jodhpur (Rajsthan). Geobios 8 (3): 119 - 122.

Kodarkar, M.S., Malu, M.B. and D.S. Dabhade (2000): Rotifer diversity in Lonar lake, an inland saline waterbody in Maharashtra, India., J. Aqua. Bio. 15 (1&2): 16 - 18.

Nayar, C.K.G. (1970): Studies on the rotifer population of two ponds at Pilani, Rajsthan. J. Zool. Surv. India, 3: 189 -190.

Patil, S.G. (1976): Plankton ecology of few fresh water bodies of Nagpur. Ph. D. Thesis, Nagpur University, Nagpur.

Patil, S.G. (1987): Plankton ecology of Gandhisagar Tank in Nagpur, India. Bull. Zoo. Surv. India. 8 (1-3): 245 - 276.

Pennak, R.W. (1978): Freshwater invertebrates of United States. John. Wiley and Sons, New York.

Somani, V.U. and M.K. Pejaver (2003): Rotifer diversity in lake Masunda, Thane, (Maharashtra). J. Aqua. Bio. 18 (2): 21 - 25.

Vankhede, G.N. and K.M. Kulkarni (1984): Physico-chemical condition and plankton populations of two lakes near Amravati, Maharashtra. Environment and Ecology, 2 (2): 132 - 133.

2013, Impact of Global Climate Change on Earth Ecosystems *Pages 187–192*
Editors: **D.R. Khanna, A.K. Chopra, Gagan Matta, Vikas Singh & Rakesh Bhutiani**
Published by: **BIOTECH BOOKS, NEW DELHI**

Chapter 24

Conservation of Spice Biodiversity to Mitigate the Climate Change Effect

Suchand Datta

Department of Vegetable and Spice Crops,
Uttar Banga Krishi Viswavidyalaya,
P.O. Pundibari, Cooch Behar – 736 165, West Bengal

India is the largest producer, consumer and exporter of spices in the World. Wild relatives of certain spice crops are at risk of extinction, threatening a valuable source of genes that are necessary to boost the ability of cultivated crops to resist biotic and abiotic stresses. The reliance on wild relatives to improve their cultivated cousins on the farm is expected to intensify as climate change makes it too hot, too cold, too wet or too dry for many existing crop varieties to continue producing at their current levels. There is an urgent need to collect and store the seeds/ planting materials of traditional cultivars and wild to mitigate the threats of climate change. The national repository of spice germplasm maintained in *ex situ* and *in situ* at IISR, Calicut conservatories are enriched regularly by undertaking collection of diverse spice crops. Apart from IISR, different AICRP centre on Spices, SAU are also engaged in collection, evaluation and maintenance of different spice germplasm. Some of the valuable collections in the germplasm like *Piper barberi* and *Piper colubrinum* are endangered species. *Piper colubrinum* is resistant to Phytophthora, Pollu beetle and *Radopholus similes*. Natural katte resistant lines of cardamom, king clove, were also collected and conserved in different conservative. The diverse nature of certain spices resulted in release of improve varieties for commercial cultivation and for combating the negative effect of climate change.

The diversity of spice crop can be utilized for identification and exploitation of novel genes for better survival of these crops under changed climate.

Keywords: *Biodiversity, Climate change, Conservation, Spice, Wild relatives.*

Introduction

Spices are defined as a strongly flavored or aromatic substance of vegetable origin, commonly used as food adjunct. Apart from used as spice most spices are used for medicinal purposes, income generation and poverty alleviation programmes in the rural areas. Spices and herbs are good not only for our taste but also for our health. They supply calcium, iron, vitamin B, vitamin C, carotene and other antioxidants (Rathore and Shekhawat, 2008). India is the largest producer, consumer and exporter of spices in the World. During 2010-11 India exported 525750 tones of spices and spice products and earned a tune of Rs 6840.70 crores. Because of the varying climates in India from tropical to sub tropical, temperate almost all spices are grown in almost all states and union territories of India. There are over 120 spices grown in different parts of the world and around 63 spices are grown in India. Threatening of drought, flood, rising temperatures and new pest and diseases are putting agricultural systems under greater pressure around the world. To survive, rural communities must diversify their activities and take full advantage of the diversity that exists in crops that can cope with the changing conditions. Loss of such diversity or genetic choices diminishes farmers' capacities to cope with extreme temperatures, drought and salinity. Thus, in the face of constant vulnerability to loss of key stress resistant types, the management of appropriate diversity of a specific crop constitutes a significant component of livelihood strategies of spice farmers in highly stressed production environments. Wild relatives of certain spice crops are on the verge of extinction, threatening a valuable source of genes that are necessary to boost up the ability of cultivated crops to resist biotic and abiotic stresses. The diverse nature of certain spices resulted in release of improve varieties for commercial cultivation and for combating the negative effect of climate change. The diversity of spice crop can be utilized for identification and exploitation of novel genes for better survival of these crops under changed climate.

Consequences of Climate Change

Climate change is predicted to cause an increase in average air temperature of between 1.4°C and 5.8°C, increases in atmospheric CO_2 concentration, and significant changes in rainfall pattern (Houghton *et al.*, 2001). Impact of climate change on four sectors of the economy, namely Agriculture, Water, Natural Ecosystems and Biodiversity and Health in four climate sensitive regions of India, namely the Himalayan region, the Western Ghats, the Coastal Area and the North-Eastern Region are more pronounced. The maximum increase in temperature will be observed in the Himalayan region. Sea level along the Indian coast is likely to rise in consonance with the global sea level rise in the future. Water yield is projected to increase in the Himalayan region in 2030s by 5-20 per cent, however, water yields are likely to be variable across the North Eastern region, Western Ghats, and Coastal region. Moderate to extreme drought severity is projected in 2030s for the Himalayan region, as compared to the other regions.

Implication of Climate Change

The impacts of shifting climate patterns on natural resources, such as water, fisheries, and forests, and on planning should be studied thoroughly for improved management of these resources to meet the needs of growing populations as the climate changes. Indian Gene centre is endowed with a number of endemic gene floras thus having a rich genetic heritage. With increasing human pressure and loss of natural vegetation, many of these spice species are already under threat. Climate change is perhaps the most important factor which influences vegetation patterns globally and is likely to significantly impact the forest ecosystems including the spice crop populations. Agro biodiversity is the feedstock for climate resilient agriculture. A major reason attributed to this is the vulnerability of the modern varieties to withstand pests and diseases and the climate change impacts. It is widely accepted now that every species and variety matters to counteract the diseases/pests threats and deleterious effect of climate change. A study conducted at IISR, Calicut using GIS models have shown that many areas that are presently suitable for spices would become unsuitable in another 25 yeas. Similarly, there would be new areas for cultivation of spices which are presently unsuitable (Datta, 2010).

Biodiversity in Spices

Agro biodiversity is the foundation of sustainable agriculture development in India. The reliance on wild relatives to improve their cultivated cousins on the farm is expected to intensify as climate change makes it too hot, too cold, too wet or too dry for many existing crop varieties to continue producing at their current levels. Large scale adoption of few improved varieties has resulted in displacing diverse genetic variability. Also the traditional knowledge associated with the use of old varieties/ land races of spices, has largely been ignored and rather disappearing. There is an urgent need to collect and store the seeds/planting materials of traditional cultivars, obsolete cultivars and wild to mitigate the threats of climate change. The Indian Institute of Spices Research, Calicut has played a phenomenal role in collecting and conserving the genetic resources of spices, which include cultivated, wild, hybrids and several endangered species. The national repository of spice germplasm maintained in ex situ and in situ conservatories are enriched regularly by undertaking collection, survey works in primary and secondary centres of origin.

Black Pepper

Black pepper (*Piper nigrum* L.), belongs to the family piperaceae is one of the most important spices and India (Western Ghats) is the centre of origin and diversity for *Piper nigrum*. The major center of diversity of the genus *Piper* is central and northern South America, where 60 per cent of the species are distributed. The genus *Piper* has more than 1000 species of which 110 species are of Indian origin. In India, the Western Ghats and the North Eastern Region are considered as two independent centers of diversity. The Indian Institute of Spices Research (IISR), Kozhikode is the National Repository for black pepper germplasm and it has a collection of 1286 wild accessions, 1300 cultivar accessions and 9 exotic collections in the gene bank. Among these valuable germplasm *Piper barberi* and *P. arboretum* are endangered and species like. *Piper colubrinum*, a source of resistance against *Phytophthora*, Pollu beetle and *Radopholus similes*. Some high yielding varieties can grow under changed climatic conditions like can tolerate adverse climatic conditions (Panniyur - 4 and Panniyur – 5), drought (Srekara and Subhakara), nematode infestation (Pournami), and *Phytophthora capsaicin* infestation (IISR Shakti).

Cardamom

Cardamom (*Elettaria cardamomum*. Maton) belongs to the family Zingiberacea, is indigenous to South India and Sri Lanka and grows either wild or semi wild condition in the tropical rain forests. Based on adaptabilty and types of panicle, three distinct types are distinguishable *viz.* Malabar, Mysore and Vazhuka. The collected germplasm are maintained at the IISR Cardamomom Research Center, Appangala, Karnataka (416 germplasm), Indian Cardamom Research Institute, Myladumpara, Kerala (612 germplasm), Cardamom Reasearch Stataion, Pampadumpara, Kerala (152 germplasm) and in Regional Research Stataion, Mudigere, Karnataka (161 germplasm). The major reasons for genetic erosion are environmental hazards, deforestation, forest fires, incidence of devastating diseases and pests and large scale cultivation of high yielding varieties (Parthasarathi, *et al.*, 2010). So, conservation of bio-diversity and sbsequent crop improvement programmes in cardamom can develop the suitable varieties which can withstand under changed climatic condition, and natural katte resistant lines of cardamom.

Ginger

Ginger (*Zingiber officinale*.Rosc) is believed to be originated in the South East Asia. India is the major producer and exporter of ginger. Eight species of *Zingiber* are found growing in Western Ghats and adjoining areas. At present, about 50 ginger cultivars, possessing various quality attributes and yield potential are prevalent in India. Besides ginger which is used as a spice as well as medicine the other economically important species of the genus are *Z zerumbet* and *Z. casumunnar*. The *ex-situ* gene bank of ginger at IISR holds 659 accessions including indigenous and exotic collection and related taxa. Certain varieties are less susceptible to rhizome rot disease (Himgiri), resistant to nematode *Meloidogyne incognita* and *M. javonica* (IISR Mahima).

Turmeric

The genus Curcuma belongs to the family Zingiberaceae and consists of about 117 species. Around 40 species were reported from India (Velayudhan *et al.*, 1999). Curcuma species differ in floral characters, aerial morphology and chemical traits (Valton, 1918; Velayudhan *et al.*, 1999). Species such as *C. aromatica* (*Kasturi manjal*), *C. caesia* (black turmeric) *C. amada*, *C. zedoaria*, *C. purpurescens*, *C. mangga*, *C. heyneana*,

C. xanthorrhiza, C. aeruginosa, C. phaeocaulis and C. petiolata are also cultivated in different places and regions. The *ex-situ* gene bank of ginger at IISR conserved 1040 accessions including indigenous and exotic collection and related taxa. NBPGR, Regional research station, Trichur holds 650 accessions. Certain turmeric cultivars are turmeric varieties are suitable for drought prone areas (CO-1 and BSR-1), hilly areas (CO-1, Roma), while others are suitable for low lying areas (Ranga) and late planting (Roma and Ranga). Besides some insects are resistant to scale insect (BSR-2 and Ranga), leaf blotch dyisease (Ranga) while others are tolerant to leaf blotch disese (IISR Alleppy, IISR Kedaram, Sona and Varna) and rhizome rot (Suguna).

Chilli

Chilies are usually grown in warm to hot and humid climate in India. Due to the long history of cultivation, outcrossing nature and popularity of the crop, large genetic diversity including local landraces have evolved. The important Capsicum species grown in India are *Capsicum annum* L., *C. annum* L. var. *avicular C. annum* var. *grossum* Sendt., *C. annum* var. *longum* Sendt, *C. chinense* Jacq, *C. frutescens, C. eximium,* and *C. pubescens. C. minimum* Roxb. Syn. *C. fastigiatum* Bhumme A large number of varieties are resistant/tolerant to different biotic factors. Cultivars like Pusa Sadabahar, Pant C-1, Pant C-2, Punjab Lal, Pusa Jwala, Jawahar -218 are resistant/tolerant leaf curl virus disease; CA-219 and CA -33 resistant/tolerant to bacterial wilt, Jawahar -218 and PBC-36 are resistant/tolerant to fruit rot, Andhra Joyti and Bahskar are resistant/ tolerant to thrips and mite K-2 for root knot nematode.

Nutmeg

Nutmeg and mace are the two distinctively different spices produced from the fruits of *Myristica fragrans*. Houtt. It was introduced to India during the 18th century. At present the crop is very popular in Kerala, Tamil Nadu and Karnataka States. About 12 species of Myristica occurs in Indo Malayan region (Krishnamoorthy and Rema 1994). The nutmeg germplasm conservatory at IISR, Kozhikode consists of 484 accessions including 406 accessions of M. fragrans and 18 accessions of related taxa. IISR Viswasree is an improved variety released from the institute recently.

Clove

The clove (*Syzygium aromaticum*.L) Merr.& Parry, belongs to the family Myrtaceae, is believed to be indigenous to the Moluccas Islands. In India, the cultivation of clove is largely restricted to South India. The germplasm conservatory of IISR, Kozhikode holds 233 accessions including two exotic collections, one each from Zanzibar and Sri Lanka.

Cinnamon

Cinnamon represents the dried inner bark of *Cinnamomum veerum* Presl of the family Lauraceae. In addition to the true cinnamon (*Cinnamomum veerum*) the other economically important species are Chinese cassia (*C.cassia* Bercht. and Presel)), Indonesian cassia (*C. burmannii*) Saigon cassia (*C. loureirii* Nees), Camphor (*C. camphora*) and Indian cassia (*C.tamala* Nees) are reported to be grown in India. The tree spices germplasm conservatory at IISR, Kozhikode has 408 accessions of *C. verum* and 72 accessions of related taxa.

Vanilla

Vanilla (*V. planifolia* Andr. Syn. *V. fragrans* Salisb.) is a climbing orchid native to tropical America and was introduced to India during the nineteenth century. Apart from *V. planifolia* other cultivated species are *V. pompona* sch. and *V. tahitensis. V. walkeriae* Wt. and *V. wightiana* Lndl. are two wild species of vanilla occurring in Western Ghats besides *V. vatsalae* and *V. pilifera* were reported from North East India. At IISR, Kozhikode 82 accessions of vanilla are conserved in addition to seedling progenies and somaclones.

Conclusion

Considerable diversity exists among the different spices species including variation in plant type, morphological and physiological characteristics, reactions to diseases and pests, adaptability and distribution in India. The establishment of NBPGR, ICAR, BSI, and various Agriculture Universities has made tremendous impacts in collection, evaluation, conservation and utilization of regional germplasm for development of spice varieties in this region. The involvement of communities in the respective areas was seen to be critical in maintaining the plots, in participatory action research, and in carrying out observations and sharing perceptions. Promotion of *in situ* and *ex situ* conservation of important species of spices in their natural habitat is urgently necessary. Adaptation to climate change is also critical for sustaining food security of the country. The diverse nature of certain spices resulted in release of improve varieties for commercial cultivation and for combating the negative effect of climate change. The diversity of spice crop can be utilized for identification and exploitation of novel genes for better survival of these crops under changed climate.

References

Datta, S. (2010) Impact of climate change in Indian horticulture. Paper presented in the International Seminar on Climate Change and Environmental Challenges of 21st Century, December 7-9, 2010 at Rajshahi, Bangladeh

Houghton, J., Ding, Y., Griggs, D., Noguer, M., Van der Linden, P. (eds.). (2001). *Climate Change 2001: The Scientific Basis. Published for the Intergovernmental Panel on Climate Change.* Cambridge University Press, Cambridge, UK and New York. pp.881

Parthasarathi, V. A., Saji, K. V. and Utpala, P. (2010) Biodiversity to enhance export of spices. *Indian Horticlture,* **55**(3): 57-61

Rathore, M.S. and Shekhawat, N.S. (2008) Incredible Spices of India: from Traditions to Cuisine. *American Eurasian Journal of Botany,* **1** (3): 85-89, 2008

Valeton, Th., 1918. New notes on the Zingiberaceae of Java and Malaya. *Bull. Jard. Bot. Buitenz. Ser. II,* **27**: 1-166.

Velayudhan K.C., Muralidharan, V.K., Amalraj, V.A.,Gautham, P.L., Mandal, S and Dineshkumar 1999. *Curcuma Genetic Resources.* Scientific Monograph, No. 4, National Bureau of Plant Genetic Resources, New Delhi, India, P-149.

2013, Impact of Global Climate Change on Earth Ecosystems *Pages* ***193–203***
Editors: **D.R. Khanna, A.K. Chopra, Gagan Matta, Vikas Singh & Rakesh Bhutiani**
Published by: **BIOTECH BOOKS, NEW DELHI**

Chapter 25

Studies on Antioxidant Activity, Total Phenolic and Flavanoid Contents of Leaf Extracts in *Thuja orientalis*

Pooja Saharan and Joginder Singh Duhan
Department of Biotechnology,
Ch. Devi Lal University, Sirsa – 125 055, Haryana

Various extracts of leaf of *Thuja orientalis* (collected from locality of Sirsa in India) were analyzed for their antioxidant activity by DPPH assay, chelating effects on ferrous ions and reducing power assay. These extracts were also screened for total phenol and flavanoid contents. Gallic acid and quercetin were used as standard for determination of total phenol and flavanoid content, respectively. Methanol and acetone extracts of leaf showed its maximum H^+ donor ability of 63.60 and 56.60 per cent at a concentration of 850μg/ml in DPPH assay. While, the same capability of ethyl acetate extract of leaf was found to be 48.62 per cent. The IC_{50} values of leaf extracts were found to be in order of methanol extract (615.62 μg/ml) > acetone extract (737.50 μg/ml) >ethyl acetate extract (850μg/ml). Almost same trend was observed for metal chelating activity of different extract of leaf. The percent reduction potential of acetone and methanol extract of leaf at a concentration of 300 μg/ml was 0.978 per cent and 0.952 per cent respectively, which is comparable with reduction potential of standard *i.e.* BHT (0.947 per cent). The order of reducing potential of various extracts of leaf was found to be in order of acetone > methanol > ethyl acetate. The total phenolic content was found to be highest in acetone extract of leaf (*i.e.* 102.25 mg) followed by methanol extract of leaf (*i.e.* 100.25 mg) where as flavanoid contents seems to be quite low in all the extracts. Hence, from this study we can conclude that methanol and acetone extracts of leaf can be used as potent antioxidants. Studies are in progress to find out the active principal responsible for this activity.

Keywords: Antioxidant activity, phenolics, flavanoids, *Thuja orientalis*

Introduction

Oxidative process is one of the most important routes for producing free radicals in foods, drugs and even in living systems (Halliwell, 1994). This oxidative damage caused by free radical is related to pathogenesis of many chronic degenerative diseases like cancer, diabetes, neurodegenerative disease, antherosclerosis, cirrhosis, malaria and AIDS (Azizova, 2002; Quinteo *et al.*, 2006; Sian, 2003; Nagler *et al.*, 2006). Reactive oxygen species (ROS), including superoxide free radical, hydrogen peroxide, hydroxyl free radical and singlet oxygen play a key role in the oxidative damage of these diseases. The most effective path to eliminate and diminish the action of free radicals which cause the oxidative stress is antioxidative defense mechanisms. Plants and herbs are an excellent source of phenolic compounds (flavonoids, phenolic acid, alcohols, stilbenes, tocopherols and tocotrienols), ascorbic acid and carotenoids which have been reported to show good antioxidant activity (Zheng and Wang, 2001). Many studies described that phenolic compounds have protective role in the oxidation of low-density lipoproteins (Andrikopoulos *et al.*, 2002) and in oxidative alterations, due to free radical and other reactive species (Soler-Rivas *et al.*, 2000).Antioxidants have been widely used as food additives to provide protection against oxidative degradation of foods (Dorman and Deans 2000; Biovati *et al.*, 2004; Hsieh *et al.*, 2001; Unver *et al.*, 2009). The uses of natural antioxidants from plant extracts have experience growing interest due to some human health professionals and consumer's concern about the safety of synthetic antioxidants in foods (Suhaj, 2006; Sun and Ho, 2005). The present investigation involves the study of antioxidant activity, total phenolic and flavanoid contents of leaf extracts of *Thuja orientalis*.

Materials and Methods

Chemicals

The organic solvents (methanol, acetone, ethyl acetate, chloroform, and hexane) were purchased from Qualigens. 1, 1-Diphenyl-2-picrylhydrazyl (DPPH), was obtained from Sigma Aldrich, USA. All other chemicals used like tris HCl, ferrozine, ferric chloride, potassium ferricyanide, trichloroacetic acid, aluminum nitrate, potassium acetate, sodium carbonate, BHT (butylated hydroxytoluene), gallic acid, Folin reagent, L-ascorbic acid, quercetin and other solvents were procured from CDH and were of analytical grade.

Preparation of Extracts

The leaves of *Thuja orientalis* was collected, washed and dried at room temperature. After grinding to the fine powder it was extracted with different solvents in increasing order of solvent polarity *viz.* hexane, chloroform, ethyl acetate, acetone, methanol and water (Flow Chart 25.1). In this method of extraction, 100 g of leaf powder was soaked for 24 h in each solvent and after recovering the supernatant, the respective solvents were added twice to the residue. The three supernatants obtained with each solvent were pooled and dried in rotary vacuum evaporator. Extraction in each solvent was done thrice and the whole procedure was repeated twice (Flow Chart 25.1).

Antioxidant Testing Assays

The antioxidant activity of the extracts was determined by using standard methods.

DPPH Radical Scavenging Assay

The DPPH assay (Hsu *et al.*, 2006) is often used to evaluate the ability of antioxidants to scavenge free radicals which are known to be a major factor in biological damages caused by oxidative stress.

One mg extract powder was dissolved in 1 ml of 50 per cent ethanol solution to obtain 1000 µg/ml sample solution. This solution was serially diluted into 100, 250, 400, 550, 700, 850 µg/ml with 50 per cent ethanol. In each reaction, the solutions were mixed with 1 ml of 0.1 mM 1,1-Diphenyl-2-picrylhydrazyl (DPPH), 0.45 ml of 50 mM Tris-HCl buffer (pH 7.4) and 0.05 ml samples at room temperature for 30 min. Fifty percent ethanol solution was used as negative control. The reduction of the DPPH free radical was measured by reading the absorbance at 517 nm. DPPH is a purple-colored stable free radical; when reduced it becomes the yellow-colored diphenylpicrylhydrazine. L-ascorbic acid was used as positive control. The antioxidant activity of test samples was evaluated by calculating the percent inhibition of superoxide anion radical by applying the following formula:

$$\text{Per cent inhibition} = [(A_0 - A_1)/A_0] \times 100$$

Where A_0 was the absorbance of the control (blank, without extract) and A_1 was the absorbance of the extract. The antioxidant activity of each sample was expressed in terms of IC_{50} (micro molar concentration required to inhibit DPPH radical formation by 50 per cent), calculated from the inhibition curve.

Chelating Effect on Ferrous Ions

The chelating effect on ferrous ions was determined according to the method of Dinis *et al.*, 1994. Measurement of the rate of color reduction allows estimation of the chelating activity of the coexisting cheaters (Yamaguchi *et al.*, 2000). Ferrozine can quantitatively form complexes with Fe^{2+}. Each extract (150, 300, 450, 600, 750 and 1000 µg ml^{-1}) in methanol (2 ml) was mixed with 0.1 ml of 2 mM $FeCl_2$ and 0.2 ml of 5 mM ferrozine solutions. After reaction for 10 min, the absorbance was measured at 562 nm. A lower absorbance indicates a stronger chelating ability. BHT (butylated hydroxytoluene) was used as the control. The percentage of ferrous ion chelating effect of test samples was evaluated by calculating the percent inhibition of superoxide anion radical by applying the following formula:

$$\text{Per cent inhibition} = [(A_0 - A_1)/A_0] \times 100$$

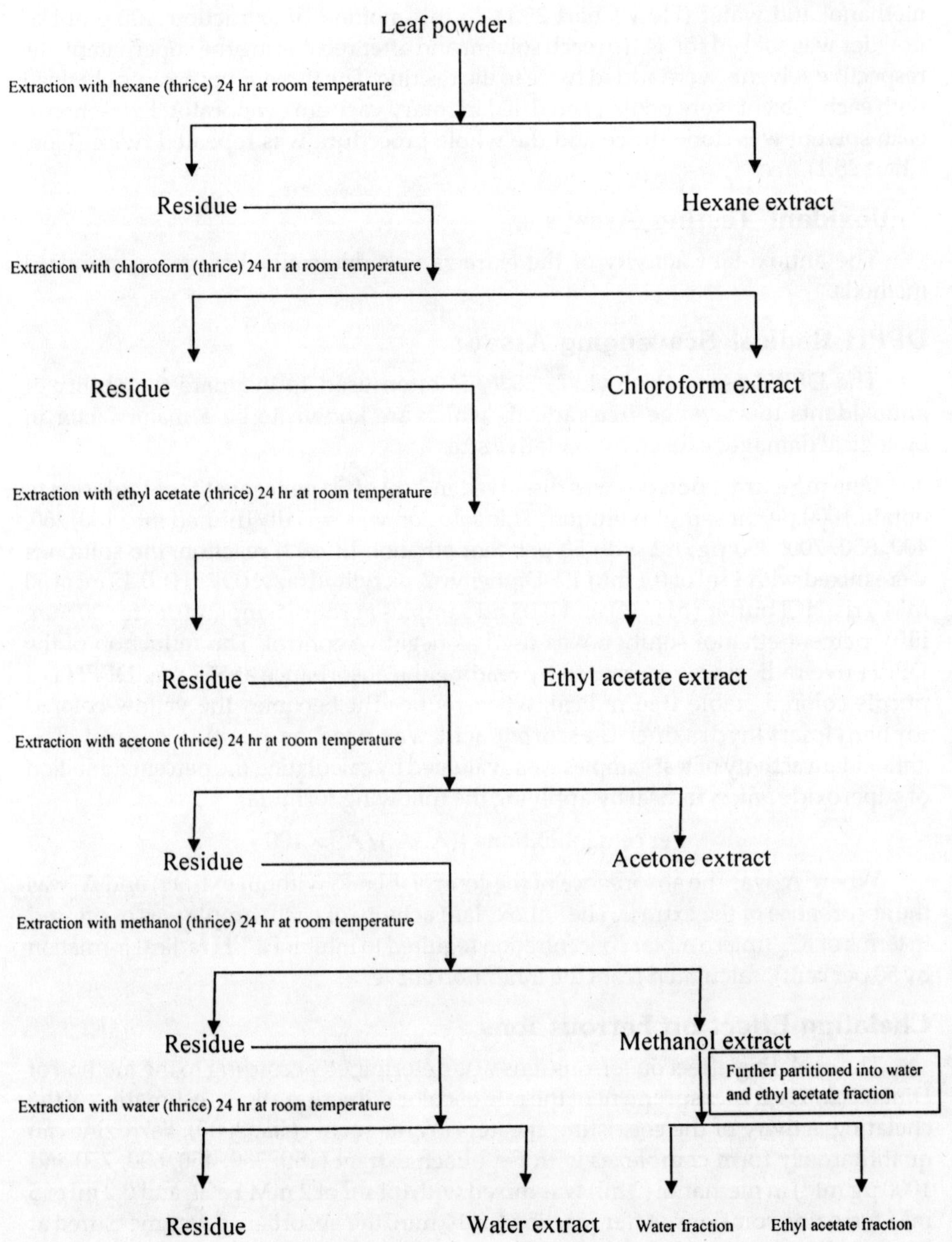

Flow Chart 25.1: Leaf Extracts of *Thuja orientalis* were Prepared by Increasing Order of Solvent Polarity

Where A_0 was the absorbance of the control (blank, without extract) and A_1 was the absorbance of the extract. The percentage of ferrous ion chelating effect of test samples were determined and compared with that of BHT (butylated hydroxytoluene), which was used as the standard (positive control).

Reducing Power Assay

The reducing power of the extracts was measured using the method described by Oyaizu (1986). Different concentrations of each extract *i.e.* 50, 100, 150, 200, 250 and 300 µg ml^{-1} in methanol, (2.5 ml) were mixed with 2.5 ml of 0.2 M sodium phosphate buffer (pH 6.6) and 2.5 ml of 1 per cent potassium ferricyanide, and the mixture was incubated at 50 °C for 20 min. Aliquots (2.5) ml of 10 per cent trichloroacetic acid (w/v) was added to the mixture which was then centrifuged at 1600g for 10 min. The upper layer (5 ml) of the solution was mixed with 5 ml of deionized water and 1 ml of 0.1 per cent ferric chloride. The mixture was shaken and left to stand for 10 min in the dark, and the absorbance was read at 700 nm in spectrophotometer. Increased absorbance of the reaction mixture indicated increased reducing power. BHT (butylated hydroxytoluene) was used as the control.

Statistical Analysis

Each value represents the mean ±SE of different measurements (not shown on the graphs).In *in vitro* studies three parallel measurements were carried out. IC_{50} values were also calculated.

Determination of Total Phenolics

Total phenolics content was determined according to the Folin-Ciocalteu method, using gallic acid as standards. Extract powders (1 mg) were dissolved in 1 ml 50 per cent methanol solution. Extract solution (0.5 ml) was mixed with 0.5 ml of 50 per cent Folin- Ciocalteu reagent. After 2-5 min, 1.0 ml of 20 per cent Na_2CO_3 was added to the mixture and incubated for 10 min at room temperature. The mixture was centrifuged at 150 g for 8 min and the absorbance of the supernatant was measured at 730 nm. The total phenolic content was expressed as gallic acid equivalents (GAE) in milligrams per gram sample.

Determination of Total Flavanoids

Sample (1 mg) was added in 1ml of 80 per cent ethanol. An aliquot of 0.5 ml was added to test tubes containing 0.1 ml of 10 per cent aluminum nitrate, 0.1 ml of 1 M potassium acetate and 4.3 ml of 80 per cent ethanol. After 40 min at room temperature, the absorbance of the supernatant was measured at 415 nm. Total flavonoid content was calculated using quercetin as standard.

Results and Discussion

DPPH Radical Scavenging Assay

DPPH is a stable nitrogen centered free radical, the color of which changes from violet to yellow upon reduction either by the process of hydrogen- or electron-donation. Specific compounds or extracts are allowed to react with the stable radical,

DPPH, in methanol solution. In the presence of hydrogen donors, DPPH is reduced and a stable free radical is formed from the scavenger. The reaction of DPPH is monitored by the decrease in absorbance of its radical at 515 nm, but upon reduction by an antioxidant, the absorption disappears (Brand- Williams *et al.*, 1995).

All the extracts of leaf exhibited potent antioxidant activity based on DPPH radical quenching assay in a dose dependent manner. Methanol and acetone extracts of leaf showed its maximum H^+ donor ability of 63.60 and 56.60 per cent at a concentration of 850µg/ml in DPPH assay (Figure 25.1). While, the same capability of ethyl acetate extract of leaf was found to be 48.62 per cent. The IC_{50} values of leaf extracts were found to be in order of methanol extract (615.62 µg/ml) > acetone extract (737.50 µg/ml) >ethyl acetate extract (850 µg/ml) as shown in the Table 25.1.

Table 25.1: IC_{50} Values of Different Extracts/Fractions of Leaf of *Thuja orientalis* in Different Antioxidant Systems

Sl.No.	Assay	Extract/Fraction/Standard	IC_{50} value (µg/ml)
1.	DPPH-radical scavenging assay	L-ascorbic acid	362.50
		Methanol	615.62
		Acetone	737.50
		Ethyl acetate	850.00
2.	Chelating effect on ferrous ions assay	BHT	475.00
		Methanol	NA
		Acetone	NA
		Ethyl acetate	NA

Table 25.2: Total Phenol and Flavonoids Contents in Different Extracts/Fractions of Leaf of *Thuja orientalis*

Sl.No.	Extract/Fractions	Phenol Content*	Flavonoids Content**
1.	Methanol	100.25	13.00
2.	Acetone	102.25	28.00
3.	Ethyl acetate	71.75	17.67

*mg gallic acid equivalent/g of extract powder, **mg quercetin equivalent/g of extract powder

In similar studies, the antioxidant properties of water, methanolic and alkaloid of *M. speciosa* leaf extracts were evaluated using the DPPH (2, 2-diphenyl-1-picrylhydrazyl) radical scavenging method. IC_{50} values of the aqueous, alkaloid and methanolic extracts were 213.4, 104.81 and 37.08g/ml, respectively. This result suggests that the methanolic extract has relatively high antioxidant activity as compared to aqueous and alkaloid extract (Parthasarathy, *et al.*, 2009).

Chelating Effects on Ferrous Ions

It was observed that methanol extract of leaf showed more metal chelating activity (47.35 per cent) than did the acetone extract (34.29 per cent) followed by ethyl acetate

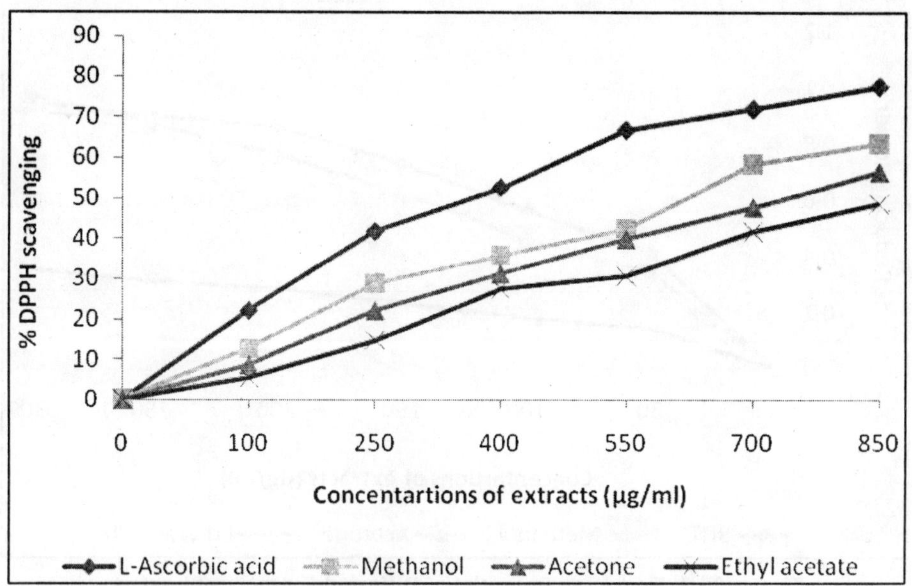

Figure 25.1: Scavenging of the DPPH Radical by Different Extracts of Leaf of *Thuja orientalis* Using DPPH Scavenging Assay

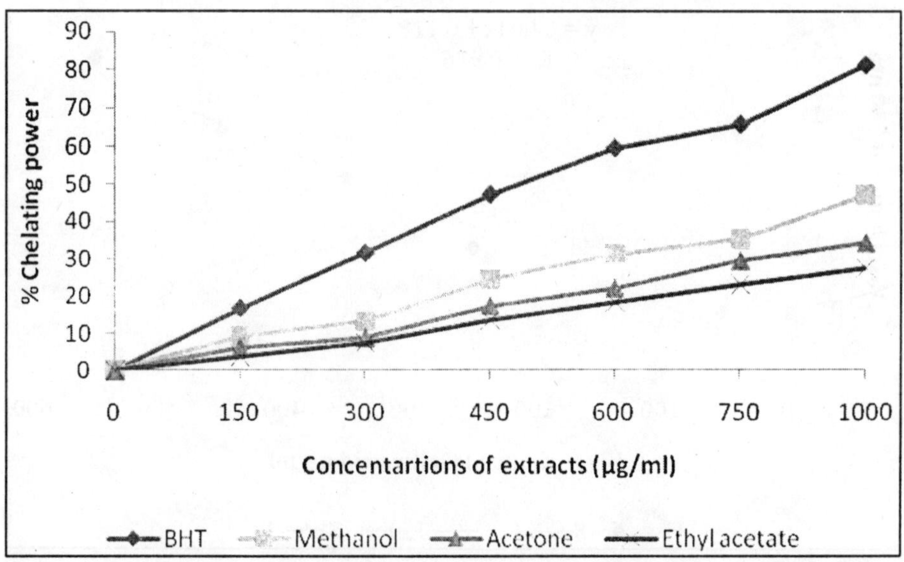

Figure 25.2: Chelating Effects of Different Extracts of Leaf of *Thuja orientalis* on Ferrous Ions Using Chelating Effect on Ferrous Ions Assay

extract (27.43 per cent) at 1000 µg/ml concentration (Figure 25.2).Ferrous chelating activity increased with increase in concentration of extract. Minimum metal chelating activity *i.e.* 3.71 per cent was observed in ethyl acetate extract at 150 µg/ml concentration.

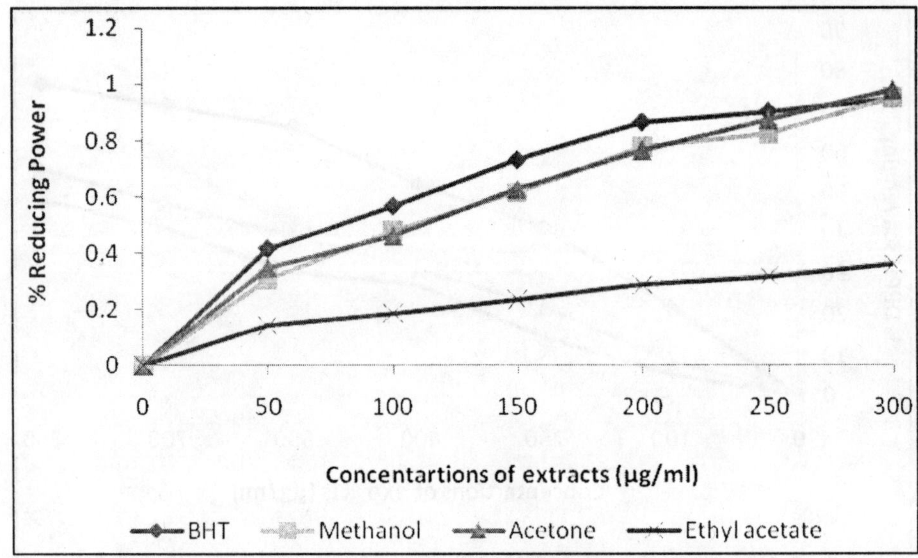

Figure 25.3: Reducing Potential of Different Extracts of Leaf of *Thuja orientalis* by Reducing Power Assay

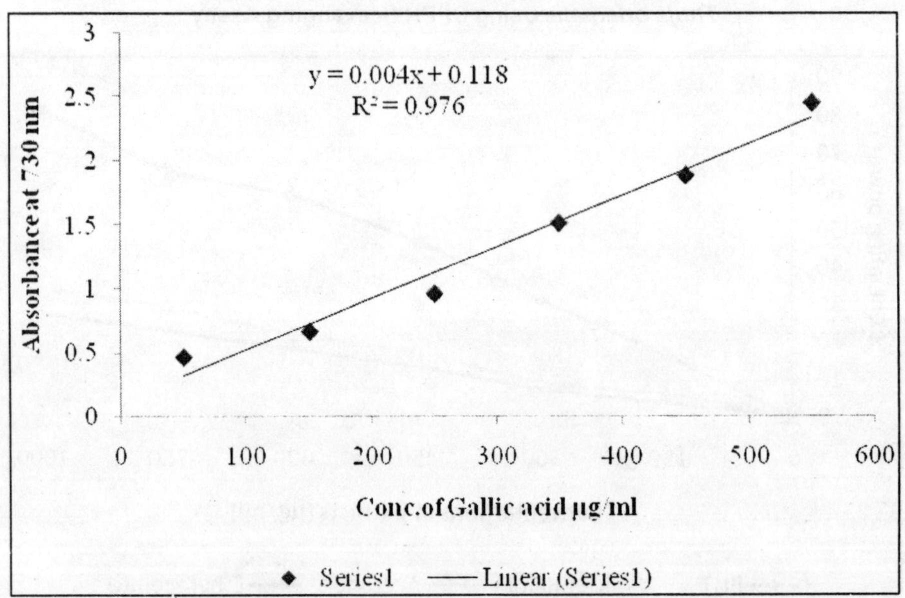

Figure 25.4: Standard Curve of Gallic Acid in Folin Ciocalteu Method for Determination of Total Phenolic Content

Our findings are compatible with the work done by Gutierrez (2010), where the efficiencies of Fe^{2+}, ferrozine complex increase with the increasing concentration of the three antioxidants *i.e.* methanol, ascorbic acid and BHT. The median inhibitory concentration (IC_{50}) values for methanol extract of *Satureja macrostema*, ascorbic acid and BHT were 66.5, 49.3 and 46.6 g/ml, respectively.

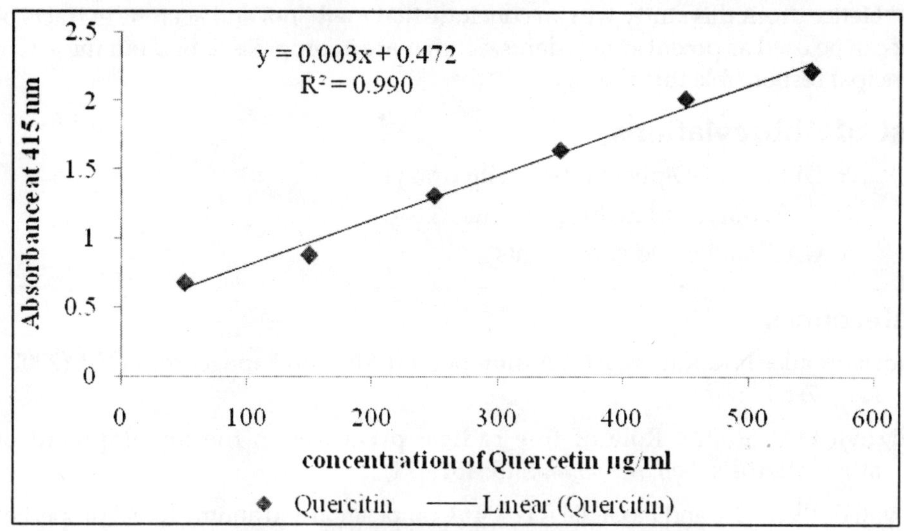

Figure 25.5: Standard Curve of Quercetin for Determination of Total Flavanoid Content

Reducing Power Assay

Overall, maximum reduction potential was exhibited by acetone extract of leaf *i.e.* 0.348, 0.462, 0.628, 0.764, 0.872 and 0.978 per cent respectively, among all the tested extracts, at all the concentrations used. The percent reduction potential of methanol extract of leaf also had better reducing power 0.952 per cent at a concentration of 300 µg/ml (Figure 25.3) which is even greater than the standard *i.e.* BHT (0.947 per cent) at same concentration. Hence this extract can be used as potent antioxidant agent. The order of reducing potential of various extracts of leaf was found to be in order of acetone > methanol > ethyl acetate.

Determination of Total Phenolics and Flavanoids

Total phenol compounds, as determined by Folin Ciocalteu method, were reported as gallic acid equivalents by reference to standard curve ($y = 0.004x$, $R^2 = 0.976$) as shown in Figure 25.4.

The total phenolic content was found to be highest in acetone extract of leaf (*i.e.* 102.25 mg) followed by methanol extract of leaf (*i.e.* 100.25 mg). Apart from that, the results also suggest that extraction by acetone could give higher phenolic content as compared to methanol and ethyl acetate (Table 25.2). These findings were likely in agreement with the finding of Sun and Ho (2005) who reported that extracting solvent significantly affected the yield of phenolic content of buckwheat extract.

The content of flavonoids was expressed in terms of quercetin equivalent (the standard curve equation: $y = 0.0033x + 0.4723$, $R^2 = 0.9907$), mg Qu./g extract (Figure 25.5).

The flavanoid content was found to be highest in acetone extract (*i.e.* 28 mg quercetin equivalent/g of extract). If we see in general the overall flavanoid content of all the extracts was found to be poor.

Hence, from this study we can conclude that methanol and acetone extracts of leaf can be used as potent antioxidants. Studies are in progress to find out the active principal responsible this activity.

List of Abbreviations

- ✰ DPPH: 1, 1-Diphenyl-2-picrylhydrazyl
- ✰ BHT: Butylated hydroxytoluene
- ✰ GAE: Gallic acid equivalents

References

Andrikopoulos NK. Kaliora AC., Assimopoulou AN. and Papageorgiou VP. (2002). *Res.*,17: 501-507.

Azizova OA. (2002). Role of free radical processes in the development of atherosclerosis. *Boil. Membrany*, 19: 451-471.

Biovati B., Ozcan M. and Piccaglia R.(2004). Composition and antimicrobial properties of *Satureja cureifolia* Ten. and *Thymbra sintenesii* Bornm. Aznav. subsp. *isaurica* P.H. Davis essential oils. *Ann. Microbiol.* 54(4):393-401.

Brand-williams W., Cuvelier ME. and Berset C. (1995).Use of a free radical method to evaluate antioxidant activity. *Lebensm. Wiss. Tech.* 28: 25-30.

Dinis TCP., Madeira VMC and Almeida LM.(1994).Action of phenolic derivatives (acetaminophen, salicylate, and 5-amino salicylate) as inhibitors of membrane lipid peroxidation and as peroxyl radical scavengers. *Archives Biochem. and Biophy.* 315: 161-169.

Dorman HJD. and Deans SG. (2000). Antimicrobial agents from plants: antibacterial activity of plant volatile oils. *J. Appl. Microbiol.* 88:308-316.

Folin O. and Ciocalteu V. (1927).Tyrosine and tryptophan determination in proteins. *J. Biology and Chemistry* 27: 627-650.

Gutierrez RMP. and Navarro YTG.(2010).Antioxidant and hepatoprotective effects of the methanol extract of the leaves of *Satureja macrostema. Pharmacognosy Magazine* 6 (22): 125-131.

Halliwell B. (1994). Free radicals, antioxidants, and human disease: curiosity, cause, or consequence. *Lancet.* 344: 721-724.

Hsieh PC., Mau JL. and Huang SH. (2001). Antimicrobial effect of various combinations of plant extracts. *Food Microbiol* 18: 35-43.

Hsu CY.(2006).Antioxidant activity of extract from *Polygonum aviculare* L. *Biol. Res.* 39:281-288.

Nagler R., Reznick A., Shafir Y. and Shehadeh N.(2006). Free radical related effects and antioxidants in saliva and serum of adolescents with type 1 diabetes metllitus. *Free Radical Res* 40: 156.

Oyaizu M. (1986). Studies on products of browning reaction prepared from glucoseamine. *Jpn. J. Nutr.* 44:307-314.

Parthasarathy S., Azizi JB., Ramanathan S., Ismail S., Sasidharan S., Said MIM. and MANSOR SM. (2009). Evaluation of antioxidant and antibacterial activities of aqueous, methanolic and alkaloid extracts from *Mitragyna speciosa* (Rubiaceae Family) leaves. *Molecules* 14: 3964-3974.

Quintero M., Brennan PA., Thomas GJ. and Moncada S. (2006).Nitric oxide is a factor in the stabilization of hypoxia inducible factor-1 alpha in cancer: Role of free radical formation. *Cancer Res.* 66: 770-774.

Sian BA.(2003).Dietary antioxidants past, presents and future? *Trends in food Sci. Technol.*,14: 93-98.

Soler-Rivas C., Espi´n JC. and Wichers HJ.(2000). *J. Sci. Food Agric.* 80: 1013-1023.

Suhaj M. (2006). Spice antioxidants isolation and their antiradical activity: a review. *J. Food Composition and Analysis* 19: 531-537.

Sun T. and Ho CT. 2005).Antioxidant activities of buckwheat extracts. *Food Chemistry* 90:743-749.

Unver A., Arslan D., Ozcan MM. and Akbulut M.(2009).Phenolic content and antioxidant activity of some species. *World Appl. Sci. J.* 6(3): 373-377.

Yamaguchi F., Ariga T., Yoshimira Y. and Nakazawa H.(2000). Antioxidant and anti-glycation of carcinol from *Garcinia indica* fruit rind. *J. Agric. Food Chem.* 48: 180-185.

Zheng W. and Wang S.(2001).Antioxidant activity and phenolic composition in selected herbs. *J. Agricultural and Food Chemistry* 49: 5165-5170.

2013, Impact of Global Climate Change on Earth Ecosystems *Pages* **205–214**
Editors: **D.R. Khanna, A.K. Chopra, Gagan Matta, Vikas Singh & Rakesh Bhutiani**
Published by: **BIOTECH BOOKS, NEW DELHI**

Chapter 26

Screening of *Aspergillus* spp. for Extra Cellular α-Amylase Activity

Ashok Kumar[1]*, Joginder Singh Duhan*[1] *and Sunil Kumar Tanwar*[2]

[1]*Department of Biotechnology,*
Ch. Devi Lal University, Sirsa – 125 055, Haryana
[2]*Department of Biotechnology and Microbiology,*
Meerut Institute of Engineering and Technology, Meerut – 250 005, U.P.

Nine fungal strains of, *A. candidus*, *A. flavus 873*, *A. flavus*, *A. terrus*, *A. ocraceous*, *A. sulphureus*, *A. oryzae*, *A. allahabadi* and *A. fumigatus* were screened for alpha-amylase activity on starch agar plates. Performance of each strain was recorded in terms of hydrolysing zone formation. On the basis of the maximum halos formation and enzyme activity by solid state fermentation, *A. sulphureus* (956 U/ml) and *A. oryzae* (940 U/ml) were selected for periodic evaluation of their dinna amylase activity. Amylase was found to be optically active at pH 8.0 and was acid stable at pH 4.5. Optimum activity was recorded at 50°C in case of *A. sulphureus* and at 40°C in case of *A. oryzae*. Thermal stability was observed in the range of 30 to 70°C for 1hour. The amylase showed maximum activity at 30°C and further increase in temperature to 70°C for 1 hour reduces the enzyme activity to 65 per cent.The amylase was slightly repressed by Na^{+} and Mg^{++} but not inhibited by Hg^{++} at 2mM concentration.$Ca^{++}$and K^{+} enhanced the amylase activity.

Keywords: Aspergillus, α-amylase, Amylase activity.

Introduction

Amylases are important enzymes which are mainly employed in the starch based industries for the hydrolysis of polysaccharides like starch into simple sugars (Apkan *et al.*, 1999; Damien *et al.*, 2010). Amylases accounts for about 30 per cent of the world's enzyme production (Rani *et al.*, 2009; Vander *et al.*, 2002). Amylases were utilized in wide range of processing industries including pharmaceutical, adhesive, sugar, textile, paper, detergent and energy manufacturing processes (Fossi *et al.*, 2005). Apart from these industries, amylases are also employed in fruit juice and brewery industries, in sewage treatment to reduce disposable solid content of sludge and for pre-treatment of animal feed to improve digestibility (Kokab *et al.*, 2003; Regulapati *et al.*, 2007; Saxena *et al.*, 2007). Amylases were produced from plants, animals, insects and micro-organisms. The amylase from the microbial sources especially from fungi (*Aspergillus* spp.) have gained much attention because of availability and high productivity of fungi (Mishra and Behra, 2008).Microbial growth and amylase production is dependent on growth condition such as type and concentration of carbon and nitrogen, metal ion requirement, pH and temperature of growth (Cherry *et al.*, 2004; Ghasemi *et al.*,2010). *Aspergillus* spp. can grow on wide range of carbon and nitrogen sources like rice bran, straws and flours of different grains and tubers like wheat, barley, corn, potato and cassava (Pandey *et al.*, 2000). Protein sources used include soyabean meal, yeast extract, peptone, NH_4Cl, $NH_4(SO_4)_2$ and meat extract (Oliveira, 2007).

Amylases were normally produced by submerged fermentation but now a days, it is produced through solid state fermentation because of lower capital cost, expenditure, cheaper fermentation media, reduced energy requirement (Ellaiah *et al.*, 2002).

In solid state fermentation usage of water is also less as compared to submerged fermentation (Pandey, 1999; Sangeeta and Rintu, 2010). Due to the ever increasing demand of alpha amylase, people are trying to increase the productivity of amylase by variety of approaches like selection of a high enzyme producing strain, process optimization usage cheaper sources and effective downstream processing (Lalit, 2010).

Therefore, the present study was undertaken to screen the indigenous efficient alpha amylase producer strains of *Aspergillus* spp. among the fungal collection of Department of Biotechnology, CDLU, Sirsa.

Materials and Methods

Chemicals

All analytical reagents and media components used were purchased from Hi-Media (Mumbai, India Pvt. Ltd).

Growth Media

For isolation of *Aspergilli*, Potato Dextrose Agar and for fungal growth and spore production was standard Czapek's Agar were used (Raper and Fennell, 1965).

Screening for Amylase Production

Preliminary screening for amylase production from *Aspergillus* isolates were carried out by starch agar plate assay on standard media (Abe *et al.*, 1988; Apkan *et al.*, 1999). After 72 hours of incubation, the inoculated plates containing media supplemented with starch were stained with Gram's iodine reagent. Plates were flooded by Iodine solution for 15 minute and washed with spring warm water to remove the excess colour.

Preparation of Spore Suspension

Ten ml distilled autoclaved water was added to the seven days old slant and the surface was gently rubbed with sterilized wire loop (Noomrio and Dahot, 1992). The stock suspension was serially diluted to prepare a conidial suspension of 1×10^6 conidia/ml with the help of spectrophotometer.

Production of Crude Amylase

The selected *Aspergillus* isolates were cultured on rice bran and soyabean basal medium as described by Apkan *et al.* (1999b). The medium consisted of rice bran (10 g) and soya bean flour (3 g) was taken in a 250 ml Erlenmeyer flask. It was moistened to 55 per cent moisture content with aqueous mineral salts solution [$MgSO_4 7H_2O$, 0.1 per cent; KH_2PO_4, 0.1 per cent; $CaCl_2$, 0.1 per cent; $FeSO_4$, 0.05 per cent and $(NH_4)_2SO_4$, 0.1 per cent]. The medium was sterilized at 121°C for 15 min. The flask containing the sterilized media were inoculated aseptically with the selected *Aspergillus* isolates having concentration of10^6 spores/ml and incubated at 30 °C for 72h in under stationary condition, the production of amylase was assayed by DNSA method (Miller, 1959).

Extraction of enzyme

The moist mouldy bran obtained after incubation was mixed with citrate phosphate buffer (pH 4.5) in the ratio 1:10 (w/v) in a 250ml Erlenmeyer flask. The mixture was shaken on an orbital shaker at 150 rpm and 28 °C for 1h. The supernatant obtained after filtration was used as source of crude enzyme.

Enzyme Assay

Amylase assay was carried out by using a reaction mixture having 1.0 ml of 1 per cent starch solution mixed with 1.0 ml of 50 mM buffer (pH 4.5) and 1.0 ml of crude enzyme, incubated for 1h at 30 °C. After incubation 2 ml DNSA reagents was added and boiled for 10 min. The development of colour due to the reaction of reducing sugar with DNSA was measured at 540 nm wavelength (Miller, 1959). One enzyme unit is equivalent to release of 1.0 μM maltose per unit time per unit volume.

The Effect of Temperature

The effect of temperature on the crude amylase activity was assayed at temperature values ranging from 25 to 90 ° C. The reaction mixture contained 1 ml of the crude enzyme, 1 ml of corn starch (1 per cent w/v) buffered with 0.1 M citrate phosphate buffer (pH 4.5) was incubated for 60 min at each chosen temperature and amylase activity was determined by the same method (Miller, 1959).

Effect of pH

Effect of pH on the crude amylase activity was determined using 0.1 per cent corn-starch suspended in buffers consisting of citrate (pH 4.0 - 6.0), phosphate (pH 6.0 - 7.5) and Tris (pH 7.5 – 9.0) at 0.1 M concentration. Each mixture was incubated at 60 °C for 1 h. Amylase activities at the different conditions studied were determined as described earlier.

The Effect of Starch Concentrations

The effect of various starch concentrations on the enzyme activity was also studied using starch solutions (1 per cent, 2 per cent, 3 per cent, 4 per cent and 5 per cent) suspended in 0.05 M NaCl in 0.1 M-phosphate buffer (pH 6.9). Each mixture was incubated at 60 °C for 1 h. Amylase activities at the different conditions studied were determined as described earlier.

Effect of Cations on Enzyme Activity

The influence of cations (K^+, Ca^{2+}, Hg^{2+}, $Mg^{2+}$and Mn^{2+}) on amylase activity was studied by adding each in the reaction solution to final concentrations of 2 mM. All metals used were in the chloride and sulphate forms. The activity was assayed under optimum conditions of both pH and temperature as required by the amylase. Activity in the absence of any additives (control) was taken as 100 per cent.

Results and Discussion

A total nine strains of *Aspergillus* spp. were screened for the production of alpha amylase by using starch agar plate method. All the isolates showed clear zone of starch hydrolysis in petridishes after iodine treatment. The preliminary amylase production analysis of all the isolates revealed that two isolate *A. oryzae* and *A. sulphureus* found to be best, showing 22mm zone of starch hydrolysis (Table 26.1). Both these isolates exhibited maximum amylase activity *i.e. A. oryzae* (940U/ml) and *A. sulphureus* (956U/ml) under solid state fermentation. These two strains were selected for further study.

Effect of Temperature on Amylase Activity

Temperature is the most important factor that affects amylase activity. The influence of temperature on amylase activity of crude amylase showed that the activity increases with the increase in temperature from 25-50 °C in *A. oryzae* and was highest at 50 °C. Further increase in temperature resulted in decrease in amylase activity (Figure 26.1). Same results were observed by Chakraborty *et al.* (2000) and Patel *et al.*, 2005. *A. sulphureus* showed optimum amylase activity at 40 °C. Above 40 °C there was reduction in amylase activity. Khoo *et al.* (1994) also reported an optimum temperature *i.e.* 40-45 °C for amylase obtained from *Aspergillus niger*.

Table 26.1: Screening of Fungal Isolates for Amylolytic Property (Amylase production)

Fungal isolates	*Width of Halos (mm)*	*Iodine Test on Broth*	*Amylase Activity (U/ml)*
Aspergillus terrus	19	++	891
Aspergillus candidus	14	+	390
Aspergillus flavus	13	+	338
Aspergillus allahabadi	15	+	405
Aspergillus flavus 873	17	+	671
Aspergillus fumigates	16	+	501
Aspergillus ocraceous	12	+	289
Aspergillus sulhureus	22	+++	956
Aspergillus oryzae	22	+++	940

+: Hydrolysed the starch and colour changes form blue to white.

Effect of pH on Amylase Activity

The crude amylase enzyme exhibited broad range of pH stability. *A. oryzae* exhibited two peaks one acidic at 4.8 pH and one basic pH 9.0.The amylase secreted by *A. sulphureus* also have pH optima at 5.6 and 8.4 pH as shown in Figure 26.2. The amylase from *A.sulphureus* and *A.oryzae* showed pH optima in alkaline range and also active at higher pH (Shandhya *et al.*, 2005).This suggest that enzyme would be useful in the process that require wide range of pH changes from slightly acidic to alkaline range or vice–versa (Abu, 2005; Yamasaki *et al.*, 1988).

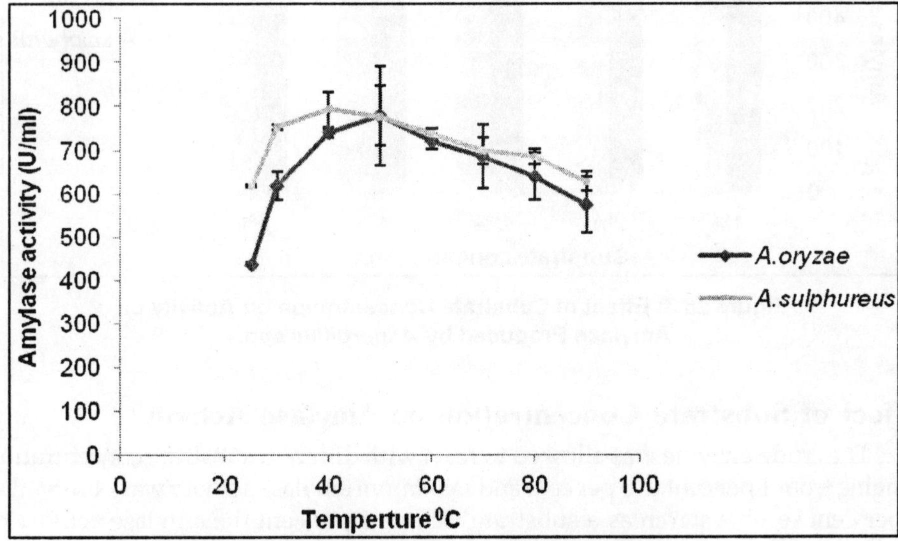

Figure 26.1: Effect of Temperature on Activity of Amylase Produced by *Aspergillus* spp.

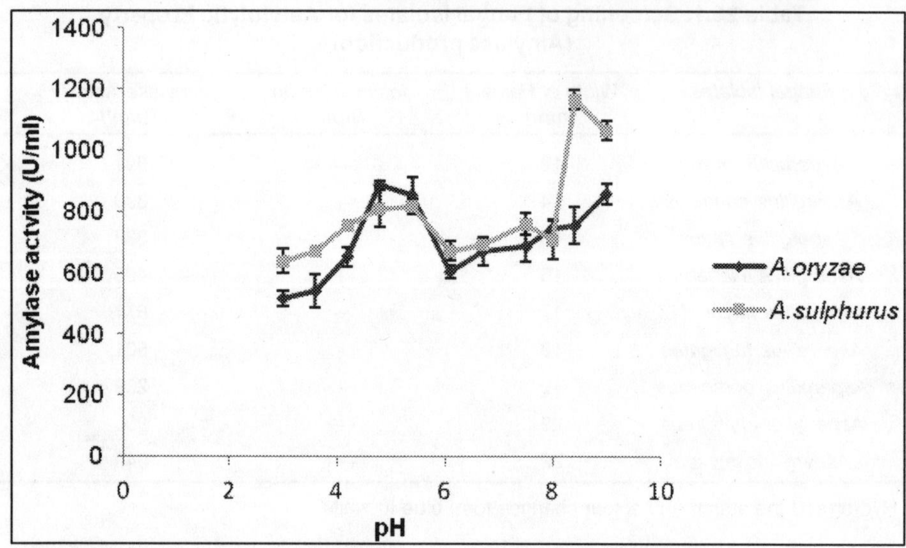

Figure 26.2: Effect of pH on Activity of Amylase Produced by *Aspergillus* spp.

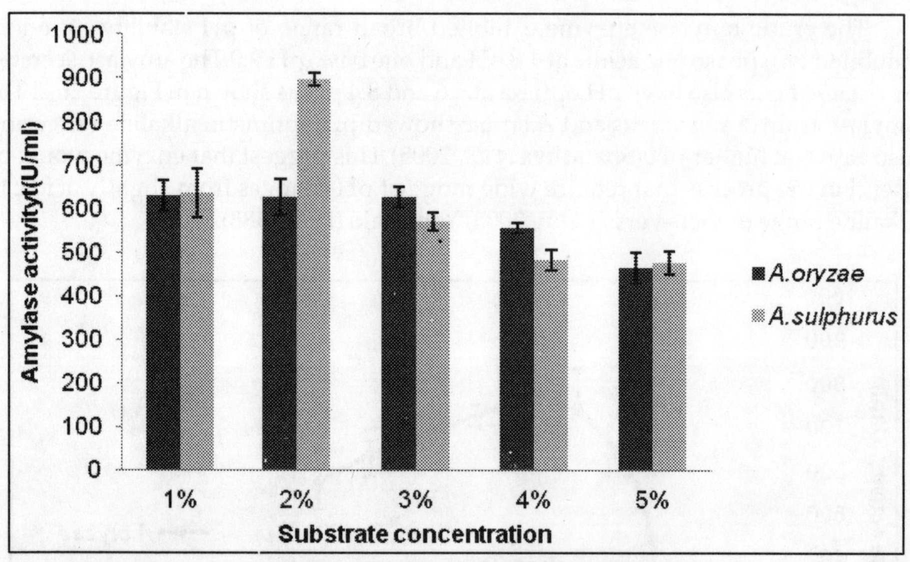

Figure 26.3: Effect of Substrate Concentration on Activity of Amylase Produced by *Aspergillus* spp.

Effect of Substrate Concentration on Amylase Activity

The crude enzyme was allowed to react with different substrate concentration ranging from 1 per cent to 5 per cent and maximum amylase activity was observed at 1 per cent soluble starch as a substrate. Above 1 per cent the amylase activity of *A.oryzae* gets decreased. The amylase activity of *A. sulphureus* was highest at 2 per cent starch concentration (Figure 26.3). These results are agree with the finding of

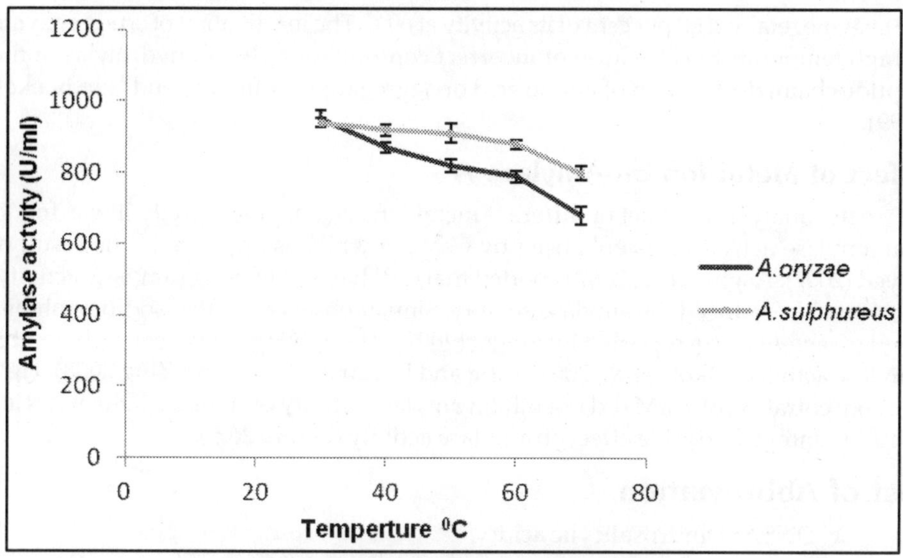

Figure 26.4: Effect of Temperature on Stability of Amylase Produced by *Aspergillus* spp.

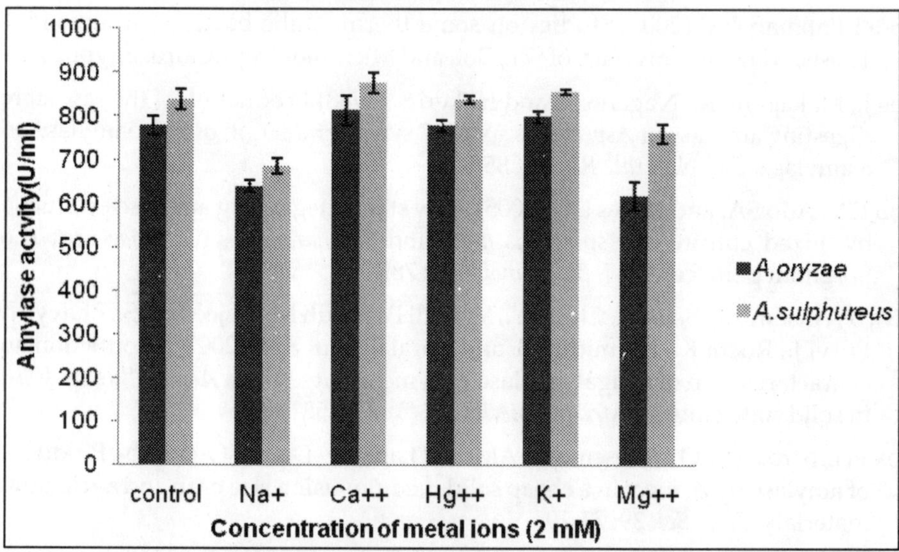

Figure 26.5: Effect of Metal Ions on Activity of Amylase Produced by *Aspergillus* spp.

Kuiper *et al.* (1998) who reported that the amylase activity was highest at 1.67 per cent of starch concentration. At low substrate concentration the active sites of enzyme are not saturated and thus the enzyme activity was increased with the increase in substrate concentration (Dixon and Webb 1997)

Thermal Stability of Amylase

Figure 26.4 showed that the enzyme from both the species showed maximum activity at 30°C and least at 70 °C. Alva *et al.* (2007) reported the same results where the enzyme retained 60 per cent of its activity at 60°C.The inactivation of crude enzyme at high temperature is because of incorrect conformation due to hydrolysis of the peptide chain, destruction of amino acid or aggregation (Schokker and Van Boekel, 1999).

Effect of Metal Ion on Amylase A

After analysis of effect of different metal ions on enzyme activity it was found that amylase activity was enhanced by Ca^{++} ion which is opposite to the result of Reyed (2007). Asgher *et al.*, 2007 reported that Ca^{++} have no effect on amylase activity. Mg^{++} and Na^{+} reduced the amylase activity. Similar observation that sodium inhibits amylase activity was reported by Reyed (2007). Though Hg^{++} have severely inhibit amylase activity (Okolo *et al.*, 2001; Yang and Liu 2004; Zhang and Zing 2008). Hg^{++} at a concentration of 2mM did not inhibit amylase activity of *A.oryzae* (Oliveria *et al.*, 2010). K^{+} showed positive effect on amylase activity (Figure 26.5).

List of Abbreviation

- ✰ DNSA: Dinitrosalicylic acid

References

Abdel-Rahman EM.(2006).Studies on some thermophilic bacterial strains. Ph.D. Thesis, Alazhar Univ. Fac. of Sci., Bot. and Microbiol.Dept.Cairo, Egypt 23-29.

Abe J., Makajoma K., Nagano H. and Hijkeri S. (1988).Production of the raw starch digesting amylase of *Aspergillus* sp. K-27 synergetic action of glucoamylase and a amylase. *Carbohydrate Res.* 75: 85-92.

Abu EA., Ado SA. and James DB. (2005). Raw starch degrading amylase production by mixed culture of *Aspergillus niger* and *Saccharomyces cerevisiae* grown on Sorghum pomace. *Afr. J. Biotechnol.* 4(8): 785-790.

Alva S., Anaemia J., Savla J., Chiu YY., Vyshali P., Shruti M., Yajeetha BS., Bhavya D., Purvi J., Ruchi K., Kumudini B. and Varalakshmi KN. (2007). Production and characterization of fungal amylase enzyme isolated from *Aspergillus* sp. JGI 12 in solid state culture. *Afri. J. Biotechnol.* 6(5): 576-581.

Apkan I., Bankole MO., Adesemowo AM. and Latunde- Dada GO. (1999b). Production of amylase by *A. niger* in a cheap solid medium using rice bran and agricultural materials. *Trop. Sci.* 39: 77-79.

Asgher M., Javaid M., Asad S., Rahman U. and Legg RL.(2007).A thermostable alpha amylase from a moderately thermophilic *Bacillus subtilis* strain for starch processing. *J. Food Engineering*. 79: 950-955.

Chakraborty K., Bhattacharyya BK. and Sen SK. (2000).Purification and characterization of a thermostable alpha amylase from *Bacillus stearothermophilus*. *Folia Microbiol.* 45: 207-210.

Cherry HM., Towhid-Hossain MD. and Anwar MN.(2004).Extracellular glucoamylase from the isolate *Aspergillus fumigatus*. *Pak. J. Biol. Sci.* 7: 1988-1992.

Damien M.,Catherine J., Patrice D. and Christopher B. (2010). Enhanced mechanical properties of partially beta-amylase trimmed starch for material application.*Carbohydrate. Poly.* 80(3): 747-752.

Dixon MW. and Webb E.C. (1997).A Textbook of Enzymes. 2nd Edn., Longmans group Ltd., London.

Ellaiah P., Adinarayana K., Bhavani Y., Padmaja P. and Srinivasulu B. (2002).Optimization of process parameters for glucoamylase production under solid-state fermentation by a newly isolated *Aspergillus* sp. *Process Biochem.* 38(4): 615-620.

Fossi B., Tavea F. and Ndjouenkeu R. (2005). Production and partial characterization of a thermostable amylase from ascomycetes yeast strain isolated from starchy soils. *Afr. J. Biotechnol.* 4: 14-18.

Ghasemi Y., Rasoul-Amini S., Ebrahiminezhad A., Zarrini G. and Kazemi A. (2010).Halo tolerant amylase production by a novel bacterial strain, *Rheinheimera aquimaris. Res. J. Microbiol.* 5: 144-149.

Khoo SL., Amirul AA., Azizan MN. and Kamaru ZM.(1994).Purification and characterization of alpha amylase of *Aspergillus flavus. Folia Microbiol.* 39: 392-398.

Kokab S., Asghar M., Rehman K., Asad M. and Adedyo O. (2003).Bio-processing of banana peel for α-amylase production by *Bacillus subtilis. Int. J. Agri. Biol.* 1: 36-39.

Kuiper J., Roels J A. and Zuidwedg M. (1998). Flow through viscometer for use in the automated determination of hydrolytic enzyme activities: Application in protease, amylase and pectinase assays. Gist-Brocades N.V. research and development,Delft, the Netherlands. *Analytical Biochem.* 90: 192-203.

Lalit S, Bharti K and Iyer LA.(2010).Purification of a bifunctional amylase/protease inhibitor from ragi (Eleusine coracana) by chromatography and its use as an affinity ligand. *J. Chromatogr.* 878(19): 1549-1554.

Miller GL.(1959).Use of dinitrosalicylic acid reagent for determination of reducing sugar. *Biotechnol. Bioeng. Symp.* 5: 193-219.

Mishra S. and Behera N. (2008).Amylase activity of a starch degrading bacteria isolated from soil receiving kitchen wastes. *Afr. J. Biotechnol.* 7: 3326-3331.

Noomrio MH. and Dahot MU. (1992). Comparative study on the production of cellulases by *Aspergillus niger* using agricultural wastes as a carbon source. *Proc. All Pak. Sci. Conf.* 1: 95-98.

Okolo BN., Ire S, Ezeogu LI., Anyanwu CU. and Odibo FJC.(2001).Purification and some properties of a novel raw starch-digesting amylase from *Aspergillus carbonarius. J. Sci. Food Agric.* 81: 329-336.

Oliveira AN.,Oliveira LA. and Andrade JS.(2010).Partial characterization of amylases of two indigenous Central Amazonian rhizobia strains. *Braz. Arch. Biol. Technol.* 53: 35-45.

Pandey A., Nigam P., Soccol C R., Soccol V T., Singh D. and Mohan R. (2000).Advances in microbial amylases. *Biotechnol. Appl. Biochem.* 31:135-152.

Pandey A., Soccol CR., Selvakumar P. and Nigam P. (1999).Solid state fermentation for the production of industrial enzymes. *Current Sci.* 77(1): 149-162.

Patel A K., Manpoothiri KM., Ramachandran S., Szakacs G. and Pandey, A. (2005).Partial Purification and characterization of alpha amylase produced by *Aspergilllus oryzae* using spent-brewing grains. *Indian J. Biotechnol.* 4: 436-341.

Rani R., Kumar A., Carlos R. and Pandey A S.(2009). Recent advances in solid state fermentation.*Biochem. Eng. J*, 44(1): 13-18.

Raper KB. and Fennell DJ.(1965).The Genus *Aspergillus*. Williams and Wilkins, Baltimore.USA686.

Oliveira A., Oliveira L., Andrade J. and Junior A.(2007).Rhizobial amylase production using various starchy substances as carbon substrates. *Braz. J. Microbiol.* 38: 208-216.

Regulapati R., Prem Malav N. and Sathyanarayana GN.(2007).Production of Thermostable α-amylases by solid state fermentation-A review. *Am. J. Food Technol.* 2: 1-11.

Sangeeta N. and Rintu B. (2009).Characterization of amylase and protease produced by *Aspergillus awamori* in a single bioreactor. *Food Res. Int.* 42(4): 443-448.

Saxena KR., Dutt K., Agarwal L. and Nayyar P. (2007).A highly and thermostable alkaline amylase from a *Bacillus* sp. PN5.*Biores. Technol.* 98: 260-265.

Schokker EP. and Van Boekel AJS. (1999).Kinetic of thermal inactivation of extracellular proteinase from *Pseudomonas fluorescens* 22F Influence of pH, Calcium and protein. *J. Agric. Food Chem.* 47: 1681 –1686.

Yamasaki Y., Tsuboi A. and Suzuki Y.(1977). Two forms of glucoamylases from *Mucor rouxianus:* properties of two glucoamylases. *Agric. Biol.Chem.* 41: 2139-2148.

Yang C. and Liu W.(2004).Purification and properties of a maltotriose-producing α-amylase from *Thermobifida fusca. Enzyme Microbial. Technol.* 35: 254-260.

Zhang J. and Zeng R.(2008). Purification and characterization of a cold-adapted α-amylase produced by *Nocardiopsis* sp. 7326 isolated from prydz bay, antarctic. Mar. *Biotechnol.* 10: 75-82.

2013, Impact of Global Climate Change on Earth Ecosystems *Pages* ***215–222***
Editors: **D.R. Khanna, A.K. Chopra, Gagan Matta, Vikas Singh & Rakesh Bhutiani**
Published by: **BIOTECH BOOKS, NEW DELHI**

Chapter 27

Extracellular Lipase of *Alternaria* spp.: Production and Properties

Pooja Jangra[1], Joginder Singh Duhan[1] and Surekha[2]
[1]Department of Biotechnology,
Ch. Devi Lal University, Sirsa – 125 055, Haryana
[2]Department of Botany,
Govt. National College, Sirsa – 125 055, Haryana

Lipases are a class of enzymes which catalyze the hydrolysis of long chain triglycerides and constitute the most important group of biocatalysts for biotechnological applications. In the present study four strains of *Alternaria* sp. were screened for maximum lipase production. Each strain was cultivated on four different media differing in their contents of mineral components, sources of carbon and nitrogen. *A. alternate* produced maximum lipase activity (914U/ml) when grown in medium III contained 0.7 per cent Na_2HPO_4, 0.1 per cent NH_4Cl, 0.002 per cent $CaCl_2.2H_2O$, 0.05 per cent NaCl, 0.025 per cent $MgSO_4.7H_2O$, 0.3 per cent KH_2PO_4 and 5.0 per cent decoction of defatted soybean flour as carbon source. The optimum pH and temperature of lipase for enzymatic activity were 7.2 and 65 °C both in *A. alternate* and *A. vitis*. The enzyme was quite stable at 40°C and pH 7.2 for 60 min in both the fungus. Further increase in temperature as well as pH, drastically reduced the stability of the enzyme. The lipase was stored for one month without any loss in activity.

Keywords: *Alternaris spp., Lipase.*

Introduction

Lipases (triacyl glycerol acylhydrolases E.C.3.1.1.3) are unique in catalyzing the hydrolysis of fats into fatty acids and glycerol at the water-lipid interface and reversing the reaction in non-aqueous media. One of the unique properties of lipases is the ability of interfacial catalysis, in which those biocatalysts become more active in presence of a substrate partially soluble in aqueous environments. Also, lipases are able to catalyze ester synthesis and transesterification inorganic media containing minute water concentration (Pandey *et al.*, 1999).

Lipases are ubiquitously produced by plants (Bhardwaj *et al.*, 2001), animals (Carriere *et al.*, 1994) and microorganisms (Olempska-Beer *et al.*, 2006). Microbial lipases are the preferred potent sources due to several industrial potentials. Many bacteria, filamentous fungi and yeasts have the ability to produce lipases (Benjamin and Pandey, 1998; Sharma *et al.*, 2001). Filamentous fungi have a greater potential for production of extracellular lipases, which are generally used as purified enzymes. Lipases from fungi were used in diverse range of industries like detergents, pharmaceuticals, beverages, dairy, etc. which make them commercially important enzyme (Hiol *et al.*, 2000; Jaeger and Reetz, 1998). Lipases produced by fungi are typically extracellular and therefore relatively easy to recover after the fermentation (Sharma *et al.*, 2001). Most of the lipase research focuses on the production of extracellular lipases through a wide variety of microorganisms. Studies on the production of extracellular lipases with *Alternaria* sp. have shown variations among different strains. However, the requirement for lipid carbon source remains essential for enzyme production. Different kinds and concentration of lipid, carbon and nitrogen sources have a significant effect on lipase yield. (Diametris *et al.*, 1996; Zarevucka *et al.*, 2003). Despite already presence of a large number of different lipolytic enzymes, the search for new lipase producer's parameters remains important due to wide range of practical and industrial applications.The knowledge and expertise in new microorganisms capable of producing lipase as well as greater understanding of their properties will be very useful in the application of such systems for some of the purposes mentioned above.

The objective of this study was to optimize and characterized the production of lipase by *Alternaria* spp.

Materials and Methods

Microorganisms and Culture Conditions

Strains of *Alternaria* spp. *i.e. Alternaria vitis* (4921), *A. solani* (4632), *A. brassicae* (5097), *A. brassicola* (1707), *A. citra* (1663), *A. alternata* (6276), and *A. cucurmeria* (6121) were taken from the microbial culture collection of IARI institute (Indian Agriculture Research Institute Pusa, New Delhi 110012). *Alternaria* spp. were grown on PDA plates at 30 °C and maintained on the slants at 4°C. The slants were sub cultured at a regular interval of three months. A spore suspension of about 10^5-10^6 spores/ml was prepared in sterilized distilled water containing about 0.01 per cent (v/v) Tween 80. This solution was used as a source of inoculum.

Chemical

All the chemicals like olive oil, decoction of defatted soybean flour, Tris-HCl, and soluble starch used in this study were procured from Hi-Media Pvt. Ltd. Mumbai. While the organic solvent *i.e.* acetone, thymolphthalein indicator, ammonium sulphate, glucose etc. used were of Qualigens Pvt. Ltd. Mumbai.

Lipase Production

Test strains of *Alternaria* sp. were cultivated in media differing in the contents of mineral components, carbon and nitrogen sources.

- ✰ **Medium I** (pH 6.0) contained 3.00 per cent peptone, 0.05 per cent $MgSO_4.7H_2O$, 0.05 per cent KCL, 0.2 per cent KH_2PO_4 and 1 per cent olive oil: glucose (0.5:0.5).
- ✰ **Medium II** (pH 7.0) contained 2.00 per cent soluble starch, 5.00 per cent corn steep liquor, 0.2 per cent KH_2PO_4 0.1 per cent $MgSO_4.7H_2O$, 0.5 per cent $CaCO_3$, 0.5 per cent soybean oil (emulsion using gum Arabic)
- ✰ **Medium III** (pH 7.0) contained 0.7 per cent Na_2HPO_4 0.1 per cent NH_4Cl, 0.002 per cent $CaCl_2.2H_2O$, 0.05 per cent NaCl, 0.025 per cent $MgSO_4.7H_2O$, 0.3 per cent KH_2PO_4, and 5.0 per cent decoction of defatted soybean flour.
- ✰ **Medium IV** (pH 4.0) contained 2.00 per cent glucose, 0.3 per cent $(NH_4)_2SO_4$, 0.2 per cent KH_2PO_4, 0.05 per cent KCL, 0.056 per cent $MgSO_4.7H_2O$.

The microorganisms were grown in 250 ml flasks containing 50 ml of culture media at 30°C under shaking condition at 200 rpm. On completion of cultivation the biomass was harvested by filtration through a Whatman filter paper no. 4 and washed thrice with distilled water then dried in an oven at 60-70 °C until complete dryness and then reweighted. The weight of biomass was calculated in grams. The content of extracellular lipase was estimated according to the lipase activity in the supernatant.

Lipase Activity

Lipase activity was determined according to the method described by Parry *et al.* (1996) using an emulsion of 10 per cent olive oil in 10 per cent gum Arabic. The emulsion was produced by treating the mixture of olive oil and gum Arabic solution in a top drive homogenizer for 10 min. The reaction mixture contained 3ml of substrate 2.5ml of deionised water, 1ml of 0.2M Tris-HCL buffer (pH 7.5) and 1.0ml lipase sample tested. The reaction was carried out at 37 °C for 2h in a shaking water bath; the reaction mixture was then supplemented with 10ml ethanol. The amount of oleic acid was determined by thymolphthalein indicator. The control sample was treated similarly using boiled enzyme samples. The lipase activity values were calculated as the average of three parallel determinations displaying a variation coefficient lowers than 5 per cent. The amount of enzyme catalyzing the formation of one micro equivalent (micromole) of oleic acid in 2h at 37°C and pH 7.5 was taken as one lipase activity unit.

$$1\,U/ml = \frac{\text{Formation of } \mu\text{mol free fatty acids}}{1 \text{ ml enzyme solution}}$$

Lipase Activity and Stability

Lipase activity was determined as mentioned before at 37°C and pH 7.5 (Parry *et al.*, 1996).

The optimum pH and the optimum temperature for lipase activity were determined by carrying out the enzymatic assay at various pH and temperature.

Lipase stability was illustrated by preincubation of lipase at different pH for pH stability or different temperature for temperature stability. The remaining activity was measured at 37 °C and pH 7.5 (Diametris *et al.*, 1996; Shafie, 2005; Suighara, 1995).

Effect of Metals on Lipase Activity

The effect of metal ions for lipase activity was also determined by using different concentrations of metals ion.

Results and Discussion

The production of lipase by the test microorganisms showed appreciable differences in their lipolytic activity.

Table 27.1 shows that all the strains of *Alternaria* exhibited lypolytic activity, however these strains differed considerably in their potentiality for lipase biosynthesis and activity levels.

Table 27.1: Production of Extracellular Lipase and Cell Biomass by Strains of *Alternaria* spp. in Different Nutrients Medium

Medium	*A. solani (4632)*		*A. alternate (6276)*		*A. brassicola (1707)*		*A. vitis (4921)*	
	LA	*DW*	*LA*	*DW*	*LA*	*DW*	*LA*	*DW*
I	131	0.697	28	2.618	154	0.862	171	2 579
II	54	0.551	217	0.371	322	0.262	45	0.401
III	247	3.905	914	4.214	657	2.339	42	2.560
IV	514	0.250	857	0.560	650	0.440	714	0.380

Incubation time: 3 days; LA: Lipase activity in U/ml; DW: Dry weight in gm.

The nutrient media composition strongly affected the biosynthesis of enzyme under study. Cultivation in media III, which contained decoction of defatted soybean flour as carbon source, showed maximum lipolytic in *A. alternate* (*i.e.* 914 U/ml), *A. brassicola* (*i.e.* 657 U/ml) as compared to the other three test media. However in media IV, which contained only glucose as carbon source, all most all the strains showed good lipase activity. Maximum activity has been shown by *A. alternate* (*i.e.* 857 U/ml) and minimum by *A. solani* (*i.e.* 514 U/ml). The order of enzyme production was *A.*

alternate > *A. vitis* > *A. brassicola* > *A. solani.* Considerable enzyme activity has been observed on medium I and II by all the four strains of fungus. No clear-cut relationship has been observed between the lipase production and biomass formation in all the four fungal strains on the four medium tested. These results are in agreed with the other results (Chander *et al.*, 1980; Chopra and Chander, 1983). One maximum and one minimum lipase producer *i.e. A. alternate* and *A. vitis* has been selected for further study.

Characterization of Lipase from *A. alternate* and *A. vitis*

Effect of pH and Temperature on Activity and Stability

As shown in Figure 27.1, the lipase activity was measured at pH ranging from 5.7 to 9.0 by using different buffers (phosphate buffer and Tris buffer).Maximum enzyme activity was observed at pH 7.2 *i.e.* 70 per cent. With the increase in pH there was drastic decrease in the enzyme activity from 70 to 26 per cent at pH 8.0 in *A. alternate.* So for the stability of lipases was concerned, the enzyme lipase was preincubated in buffer between pH 6.0 to 9.0 for 60 min. The enzyme was quite stable over a pH range of 6.0 to 7.5. Maximum stability up to 56 per cent was retained at pH 7.2. These results are similar to other reported lipases (Chander *et al.*, 1980; Davranv *et al.*, 1984; Gilber *et al.*, 1991; Suighara *et al.*, 1995) where the optimum pH for lipase activity ranged between 6.5 and 8.5.

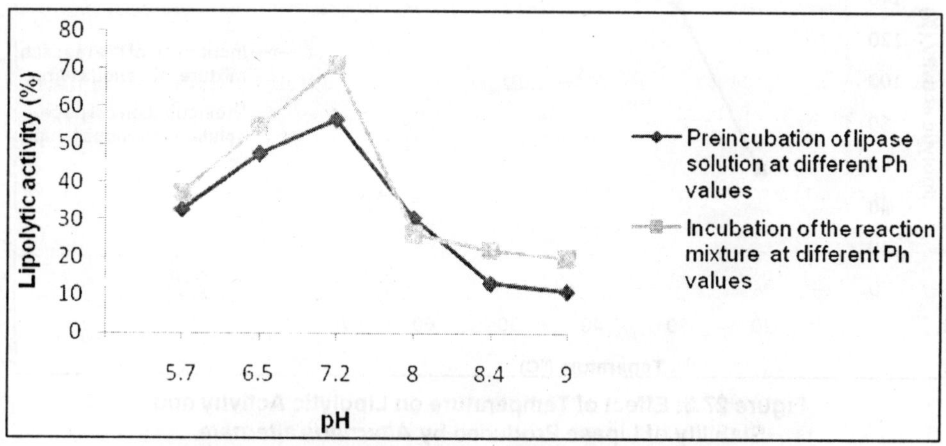

Figure 27.1: Effect of pH on Lipolytic Activity and Stability of Lipase Produced by *A. alternate*

As soon in the Figure 27.2, the maximum enzyme activity was obtained at pH 7.2 *i.e.* 58 per cent. Further increase in pH led to decrease in lipase activity which was reduced upto 20 per cent at pH 8.0 in *A.vitis*. The enzyme was quite stable at pH range of 5.8 to 7.8 and retained maximum activity at pH 7.2 *i.e.* up to 43 per cent. Diametris *et al.* (1996) reported that in *Alternaria* sp. lipase rapidly denatures at alkaline pH lose approximately 50 per cent of the initial activity after 55 min.

Figure 27.3 showed the effect of temperature on enzyme activity as well as on stability. Lipase from *A. alternate* was optimally active at a temperature of 40 °C and

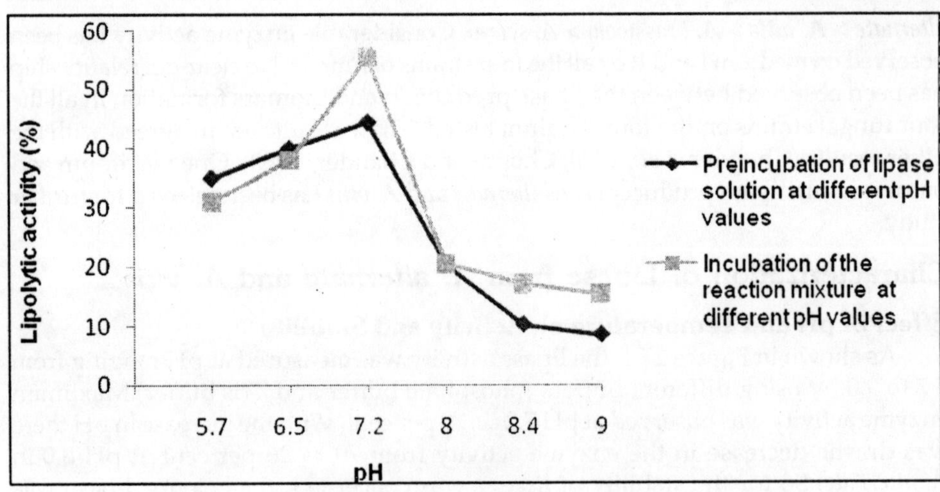

Figure 27.2: Effect of pH on Lipolytic Activity and Stability of Lipase Produced by *A. vitis*

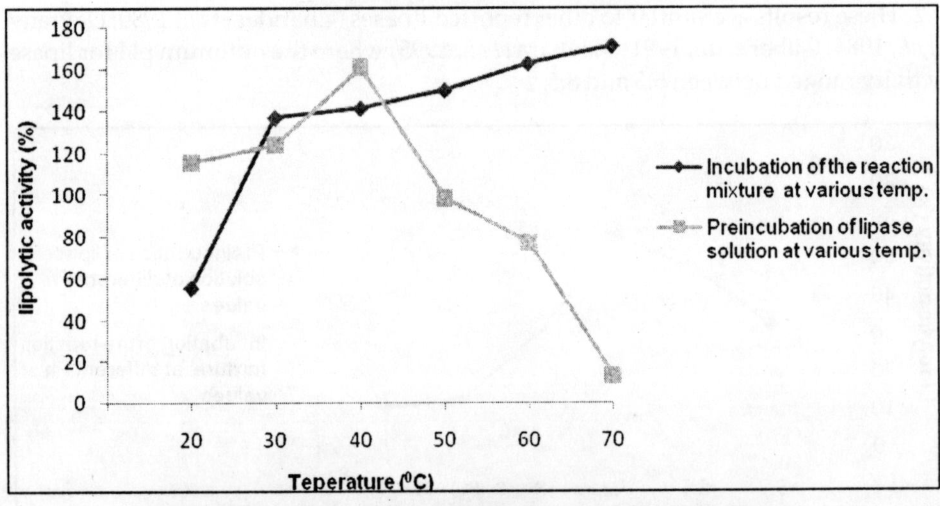

Figure 27.3: Effect of Temperature on Lipolytic Activity and Stability of Lipase Produced by *Alternaria alternate.* Enzyme activity was measured at 37°C.

its activity increases with the increase in temperature from 40 to 70 °C *i.e.* 162 per cent under incubation condition. A similar phenomenon of substrate stabilization of an enzyme at high temperature was observed in *Chromobacterium viscosum* lipase (Sugihara and Isobe, 1974) and in *staphylococcal* lipase (Stefan *et al.*, 1983).

The preincubation of lipase (without substrate) at 40 °C for 60 min caused increased lipolytic activity (*i.e.* 160 per cent) further increase in temperature reduced the stability of the enzyme which remain only 11 per cent at 70°C in *A. alternate.* Preincubation of lipase at 60 °C for 60 min caused a drastic loss of activity which reduced to 60 per cent.

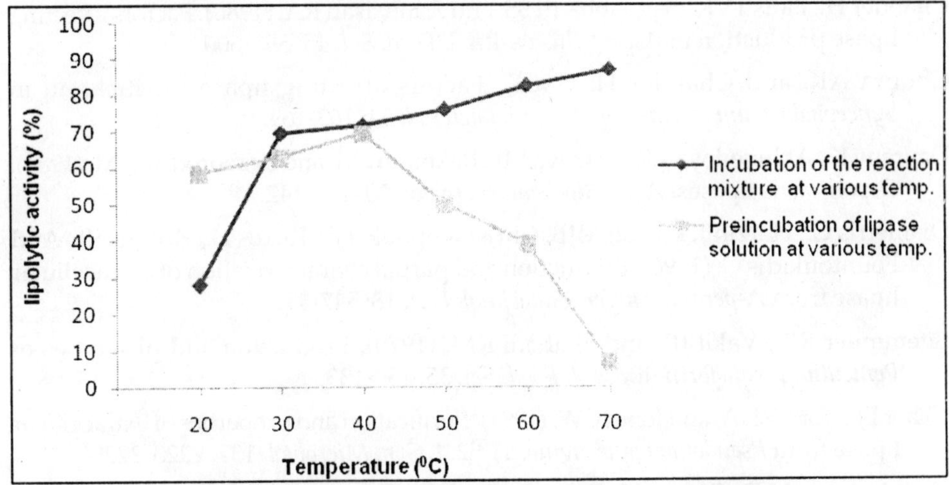

Figure 27.4: Effect of Temperature on Lipolytic Activity and Stability of Lipase Produced by *Alternaria vitis*. Enzyme activity was measured at 37 °C.

Almost similar trend of effect of temperature has been observed on both *i.e.* activity as well as stability of lipase produced by *A. vitis* Figure 27.4.

Storage Stability

Lipase preparations did not lose any activity when stored for 30 days at temperature ranging from -20 to 4 °C. Stefan *et al.* (1983) had studied the effect of time and temperature on lipase activity. They found that lipase preparation did not lose any activity when stored for 35 days at temperatures ranging from 4 to 37 °C lipases from *Penicillium requefortii* (Eitemmiler *et al.*, 1973.

Conclusion

The results of present study showed that the production of lipase from *Alternaria* sp. would be useful for food processing including flavor generation of dairy products. In relation to the different test strains, *A. alternate* was the best producer of extracellular lipase and its synthesis is induced by the presence of a lipid, carbon and nitrogen sources. This investigation suggests that various pH, temperature condition and different media can be further explored for lipase production. Effect of other possible metal ions can also be detected.

References

Benjamin S. and Pandey A. (1998). *Candida rugosa* and its lipases: A retrospect, *J. Sci. Ind. Res.*57:1–9.

Bhardwaj K., Raju A. and Rajasekharan R. (2001). Identification, purification and characterization of a thermally stable lipase from rice bran. A new member of the (phospho) lipase family. *Plant Physiol.*127: 1728–38.

Carriere. F., Thirstrup. K. and Hjorth. S. (1994). Cloning of the classical guinea pig pancreatic lipase and comparison with the lipase related protein 2. *FEBS*. Lett. 388: 63–68.

Chander H., Batish VK., Sannabhadti SS. and Srinivasan RA. (1980). Factors affecting lipase production in *Aspergillus wentii*. *J. Food. Sci.* 45: 598-600.

Chopra AK. and Chander H. (1983). Factors affecting lipase purification in *Syncephalastrum racemosum*. *J. Appl. Bacterial.* 54: 163-169.

Davranv K., Akhmedoya ZR., Rizaeva MI., Rakhimov M. and Bezborodov AM. (1984). *Osporo lactis* lipases. *Appl. Biochem. Mcrobiol.* 20: 135-142.

Diametris G., Hatzinik kolaou MJB. Christakopoulos P., Kekos D., Kolisis NF. And Fountoukidis G. (1996). Production and partial characterization of extracellular lipase from *Aspergillus niger*. *Biotechnol. Lett.* 18: 547-552.

Eitemmiler RR., Vakil JR. and Shahani KM. (1970). Production and properties of *Penicillium requefortii* lipase. *J. Food. Sci.* 35: 130-133.

Gilber EJ., Cormish A. and Jones CW. (1991). Purification and properties of extracellular lipase from *Pseudomonas aeruginosa* EF2. *J. Gen. Microbiol.* 137: 2223-2229.

Hiol, A., Jonzo M.D., Rugani N., Druet D., Sardo L. and Comeau LC. (2000). Purification and characterization of an extracellular lipase from a thermophilic *Rhizopus oryzae* strain isolated from palm fruit. *Enz. Microbiol. Technology*, 26: 421-430.

Jaeger XE. and Reetz MT. (1998). Microbial lipases from versatile tools for biotechnology. *Trends in Biotechnology*, 16: 396-403.

Olempska-Beer ZS., Merker RI. and Ditto MD. (2006).Food processing enzymes from recombinant microorganisms. *Reg. Toxicol. Pharmacol.*45:144–58.

Pandey A, Benjamin S., Soccol CR., Nigam P., Krieger N. and Soccol VT. (1999). There are of microbial lipases in biotechnology, *Biotechnol. Appl. Biochem.* 29: 119–131.

Parry RM., Chandan RC. and Shahani KM. (1996). Rapid and sensitive assay for milk lipase. *J. Dairy. Sci.* 49: 356-360.

Shafie MS. (2005). Production and partial purification of lipase from *Aspergillus niger* 599. *Union Arab Biologists* 4: 511-524.

Sharma R., Chisti Y. and Banerjee UC. (2001) Production, purification, characterization and applications of lipases, *Biotechnol. Adv.* 19: 627–662.

Stefan T., Waleria H. and Janusz J. (1983). Purification and some properties of the *Staphylococcal* extracellular lipase. *Biochemica. Biophysica. Acta.* 749 (3): 312-317.

Sugihara M. and Isobe M. (1974). Effect of temperature and state of substrate on the rate of hydrolysis of glycerides by lipase. *Biochem. Biophys. Acta.* 341: 195-200.

Suighara A., Senoo T., Enoki A. and Shimada Y. (1995). Purification and characterization of a lipase from *Pichiaburtonii*. *Appl. Microbiol. Biotechnol.* 43: 277-281.

Zarevucka M., Kejik Z., Saman D., Wimmer Z. and Haq L. (2003). Enantioselective properties of induced lipases from *Geotrichum*. *Enz. Microb. Technol.* 37: 481-486.

2013, Impact of Global Climate Change on Earth Ecosystems *Pages* ***223–240***
Editors: **D.R. Khanna, A.K. Chopra, Gagan Matta, Vikas Singh & Rakesh Bhutiani**
Published by: **BIOTECH BOOKS, NEW DELHI**

Chapter 28

Study of Avifauna Visited to Borgaon Region of Nagpur, Maharashtra, India

Kishor G. Patil and Rajendra V. Tijare
Govt. Institute of Science,
Nagpur, Maharashtra

The favorable ecological conditions like availability of food, wetlands and marshy places attracting the various birds. The present investigation carried out to document avifauna visited Borgaon area, the south-east part of Gorewada national park. The birds were recorded during June 2010 to May 2011in morning and evening hours. The birds visited to this area prominently for food and shelters were local migratory and residential.

Total 72 species having different feeding habits were detected during the observation. From the observed species some birds found in all the three seasons while others found in winter. The purple rumped sunbird and Shikra observed during rainy season only.

Keywords: Avifauna, Borgaon, Nagpur.

Introduction

Birds have great efficiency to fly. They are good bioindicators in terms of potential pollinators and scavengers. Population of birds is a sensitive indicator of pollution in both terrestrial and aquatic ecosystem (Gaston, 1975). Many ecologists made their attentions to relation of birds with the other communities. Now a day due to civilization, the birds habitats are going to destroyed, which directly affect on their reproduction and nesting.

Borgaon region of Nagpur (Maharashtra) is a fast developing area but still greeneries is present so that most of the birds are visited to this region for feeding as well as nesting. The area is adjacent to villages and containing many bushes, trees providing suitable habitat for most of the birds. Marshy land and damp places is abundant in the region and avail rich food in terms of insects, beetles, grubs, crustaceans and amphibians. An attempt has been made to compare the avian diversity with their status and feeding habitat.

Methodology

For the study of comparative avifaunal community, counting method is applied. The birds were counted with the help of binocular and snapped with Soni Handy cam 3.1 mega pixels (25 x optical zoom). The survey of avifauna from this region was undertaken during the period from June 2010 to May 2011 in the morning and evening hours only. After identification of birds checklist of bird species was prepared as per the directions by Salim Ali (1995) and Grimmet *et al.* (1998).

The birds observed in the habitat were segregated into Rare (R), Residential (RS), Resident migratory (RM), Occasional (O) and Common (C).

Table 28.1: Birds of Borgaon Region, Nagpur

Sl.No.	*Order/Family and Zoological Name*	*Common Name*	*Status*
	Order- Pelecani-formes, Family- Ardeidae		
1.	*Ardea cinerea*	Grey Heron	RM
2.	*Aredeola grayii*	Indian Pond Heron	RS, C
3.	*Bubulcus ibis*	Cattle Egret	RS, C
4.	*Casmerodius albus*	Great egret	RS, C
5.	*Ardea purpurea*	Purple Heron	RM
	Order- Gruiformes, Family- Rallidae		
6.	*Amauronis phoenicurus*	White-breasted Waterhen	RS, O
	Order- Charadriiformes, Family- Charadriidae		
7.	*Vanellus indicus*	Red wattled lapwing	RS, C
	Family- Scolopacidae		
8.	*Actitis hypoleucos*	Common Sandpiper	RS, O
	Order- Columbiformes, Family- Columbidae		
9.	*Stigmatopelia senegalensis*	Laughing Dove	RS, C
10.	*Spilopelia chinensis*	Spotted Dove	RS, O
11.	*Streptopelia decaocto*	Eurasian Collared Dove	RS, O
12.	*Columba livia*	Blue rock pigeon	RS, C

Contd...

Table 28.1–Contd...

Sl.No.	Order/Family and Zoological Name	Common Name	Status
	Order- Cuculiformes, Family- Cuculidae		
13.	*Centropus sinensis*	Greater coucal	RS, O
14.	*Centropus bengalensis*	Lesser coucal	RS, O
15.	*Cuculus canorus*	Common cuckoo	RS, C
16.	*Clamator jacobinus*	Pied Crested Cuckoo	RS, O
17.	*Hierococcyx varius*	Brain fever bird	RS, O
18.	*Eudynamys scolopacea*	Asian Koel	RS, C
	Order- Coraciiformes, Family- Alcedinidae		
19.	*Alcedo athis*	Small blue kingfisher	RS, O
	Family- Halcyonidae		
20.	*Halcyon smyrnensis*	White breasted kingfisher	RS, C
	Family- Meropidae		
21.	*Merops orientalis*	Green Bee-eater	RS, C
22.	*Merops phillippinus*	Blue tailed bee eater	RS, O
	Family- Coraciidae		
23.	*Coracias benghalensis*	Indian Roller	RS, O
	Family- Bucerotidae		
24.	*Ocyceros birostris*	Indian Grey Hornbill	RS, O
	Family- Upupidae		
25.	*Upupa epops*	Common Hoopoe	RS, O
	Order- Piciformes, Family- Megalaimidae		
26.	*Megalaima haemacephala*	Coppersmith Barbet	RS, R
	Order- Psittaciformes, Family- Psittacidae		
27.	*Psitacula krameri*	Rose Ringed Parakeet	RS, C
28.	*Psitacula eupatria*	Alexandrine Parakeet	RS, O
	Order- Passeriformes, Family- Laniidae		
29.	*Lanius schach*	Rufous-backed Shrike	RS, O
30.	*Lanius vittatus*	Bay-backed Shrike	RS, O
	Family- Dicruridae		
31.	*Dicrurus macrocercus*	Black Drongo	RS, O
	Family- Sturnidae		
32.	*Acridotheres tristis*	Indian Myna	RS, C
33.	*Sturnia pagodarum*	Brahminy starling	RS, C
34.	*Gracupica contra*	Asian pied Starling	RS, C
	Family- Pycnonotidae		
35.	*Pycnonotus cafer*	Red vented bulbul	RS, C

Contd...

Table 28.1–Contd...

Sl.No.	Order/Family and Zoological Name	Common Name	Status
	Family- Timaliidae		
36.	*Turdoides striata*	Jungle Babbler	RS, C
	Family- Sylviidae		
37.	*Chrysomma sinense*	Yellow eyed babbler	RS, O
	Family- Muscicapidae		
38.	*Saxicoloides fulicatus*	Indian Robin	RS, O
39.	*Saxicola ferreus*	Grey Bushchat	RS, R
40.	*Saxicola caprata*	Pied Bushchat	RS, R
41.	*Cercomela fusca*	Brown Rock Chat	RS, R
42.	*Ficedula nigrorufa*	Fly catcher	RS, O
43.	*Phoenicurus ochruros*	Black Redstart	RS, O
	Family- Nectariniidae		
44.	*Cinnyris asiaticus*	Purple sunbird	RS, O
45.	*Leptocoma zeylonica*	Purple rumped sunbird	RS, O
	Family- Hirundinidae		
46.	*Hirundo rustica*	Common swallow	RS, C
47.	*Hirundo smithii*	Wire-tailed Swallow	RS, O
48.	*Hirundo daurica*	Red rumped Swallow	RS, O
49.	*Hirundo concolor*	Dusky crag martin	RS, R
	Family- Estrildidae		
50.	Lonchura punctulata	Spotted munia	RS, C
51.	*Lonchura caniceps*	Grey-headed Munia	RS, O
52.	*Euodice malabarica*	White-throated Munia	RS, O
	Family- Cisticolidae		
53.	*Prinia socialis*	Ashy Prinia	RS, O
54.	*Prinia sylyatica*	Jungle prinia	RS, O
55.	*Prinia hodgsonii*	Grey-breasted Prinia	RS, O
56.	*Orthotomus sutorius*	Common tailor bird	RS, R
	Family- Passeridae		
57.	*Petronia xanthocollis*	Yellow throated sparrow	RS, O
58.	*Passer domesticus*	House sparrow	RS, C
	Family- Campephagidae		
59.	*Pericrocotus brevirostris*	Minivet	RS, R
	Family- Corvidae		
60.	*Dendrocitta vagabunda*	Indian tree pie	RS, C
61.	*Corvus splendens*	House crow	RS, C
62.	*Corvus macrorhynchos*	Jungle crow	RS, O

Contd...

Table 28.1–Contd...

Sl.No.	Order/Family and Zoological Name	Common Name	Status
	Family- Oriolidae		
63	*Oriolus oriolus*	Eurasian Golden Oriole	RS, O
	Order- Galliformes, Family- Motacillidae		
64	*Motacilla cinerea*	Grey Wagtail	RM, O
65	*Anthus rufulus*	Paddy field Pipit	RM, C
66	*Motacilla flava*	Yellow Wagtail	RS, O
	Family- Zosteropidae		
67	*Zosterops palpebrosus*	Oriental White-eye	RS, O
	Family- Acrocephalidae		
68	*Acrocephalus concinens*	Blunt-winged Warbler	RS, O
	Family- Phasianidae		
69	*Francolinus pondicerianus*	Grey Francolin	RS, R
	Order- Accipitriformes, Family- Accipitridae		
70	*Accipiter badius*	Shikra	RS, R
	Family- Campephagidae		
71	*Milvus migrans*	Black kite	RS, R
72	*Elanus axillaris*	Black-shouldered Kite	RS, R

Result and Discussion

During the present investigation 72 bird species from 12 orders and 35 families were recorded. Maximum species were observed during monsoon and winter months in compare to summer months probably due to scarcity of food.

In the present investigations birds such as Wagtails, Cuckoos, Bee eaters, Fly catchers, Warblers, Robins, Red wattled lapwing, Drongos were recorded of which food was basically insect grubs, insects, beetles, midges, flies, grasshoppers etc. Certain picivorous birds like Kingfishers, Egrets and Pond herons bird species are also recorded during monsoon as their chief food like insects, crustaceans, tadpoles, small amphibians was abundant in the area.

White breasted waterhen was found feeding on grasses, small plants, grains, worms and insects. Non aquatic carnivorous species like kites, Shikara, Shirke were recorded to feed upon grasshopper, snails, crabs, frogs, snakes, hatchlings of birds.

Omnivorous bird species like Larks, Mynas, Babblers, Asian Koel, Red vented bulbul, Indian Hornbill and Indian tree pie also found as residential bird species. Their chief food was consists of grains, fruits and small invertebrates. Sparrows and Bush chats were very common and observed frequently. Presence of these birds may due to large number of hedge plant present in this area.

In winter season when there was flowering and fruiting frugivorous species *i.e.* common green pegion, Parakeets, Coppersmith barbet were observed to be feeding on various fruits like guava, custard apple, berries, wild fig and their fruit buds.

Birds of Borgaon Region, Nagpur

White-breasted Waterhen

Red Wattled Lapwing

Laughing Dove

Greater Coucal

White Breasted Kingfisher

Green Bee-eater

Indian Grey Hornbill

Rose Ringed Parakeet

Black Drongo

Red Vented Bulbul

Indian Robin

Yellow Eyed Babbler

Asian Koel (Male)

Asian Koel (Female)

Purple Rumped Sunbird

Wire-Tailed Swallow

Spotted Munia

Common Tailor Bird

Ashy Prinia

Paddy Field Pipit

Grey Francolin

Shikra

Small nectar feeder species like purple sunbird and purple rumped sunbirds were recorded while feeding on nectar and pollen grains from different varieties of flowers.

Presence of birds in this study area was observed according to availability of food in different seasons. Similar findings were recorded by C. Sivaperuman and E.A. Jayson (2000) in their respective study areas. Birds are very significant component of biodiversity and important indicator of balanced and living ecosystem population of birds in a particular ecosystem is depending on the composition of ecosystem, environmental conditions, seasonal variations and availability of food in ecosystem (Jayson, 2002). With this reference, it is concluded that visiting and residential avifauna of Borgaon region, Nagpur might be food specific.

Acknowledgment

The authors are thankful to the Director, Institute of Science, Nagpur for providing necessary facilities for conducting this work.

References

Ali, S. (1995): The book of Indian Birds, Oxford University Press, new Delhi, 13th Edn.

Ali, S., and Dillon Ripley, S. (1995): A pictorial guide to the bird of the Indian subcontinent. B.N.H.S. Mumbai and Oxford University, Mumbai.

Ali, S., and Leeq Futehally, (2004): Common Birds, National Book Trust, New Delhi, India. pp 1- 126.

Chitampalli, M. and Bhatkhande (1998): Hansdev Mrigpakshi Shastra, M.S.Board of literature and culture, Mumbai- pp 1- 279.

Gaston, A.J. (1975): Estimating bird population, Journal of the Bombay Natural History Society 72: 271 – 283.

Grimmett, R., C. Inskipp and T. Inskipp (1998): Birds of the Indian subcontinent, Oxford University Press, New Delhi.

Jayson, E.A. and D.N. Mathew (2002): Structure and composition of two bird communities in Southern Western Ghats. J.Bom.Nat.History,Society (1); 08 – 25.

Kulkarni, A.K. and Kanwate, V.S. (2010): Avifauna of forest Jaldhara, Kinwat, Dist-Nanded, M.S., J.Aqua. Biol. 25 (1); 46-51.

Kulkarni, A.K., Kanwate, V.S. and Deshpande, V.D. (2005):Birds in and around Nanded city, Maharashtra. Zoo's Print journal, 20 (11): 2076 – 2078.

Kurhade, S. (2010): Status of Diversity of Avifauna in Jaikwadi Reservoir, Maharashtra, J.Aqua. Biol. 25 (1); 32 - 40.

Mankandan, R and A. Pittie (2001): Standardized common and scientific names of birds of Indian continent. Buceros, 6: 1- 37.

Newton, P.N., Brudin, S. and Guy, J. (1986): The birds of Kanha Tiger Reserve, Madhyapradesh, India. J. Bom. Nat. His. Society 83 (3); 977 – 998.

Sivaperuman, C. and E.A.Jayson (2000): Birds of Kole Wetlands, Trissur, Kerla. Zoos. Print journal, 15 (10): 344 – 349.

Thakor, F.J., Acharya, C.A., Bhoi, D.K., Prajapati, J.R. and Vaidya, J.S. (2010): A comparative study of avifauna from two reservoirs in Kheda district, Gujrat (India); J.Aqua. Biol. 25 (1); 41 - 45.

Wadatkar, J.S. and Kasambe, R. (2002): Checklist of Birds from Pohara, Dist.- Amravati, Maharashtra, Zoo's Print Journal, 17 (66); 807 – 811.

Yahya, H.S.A. (1988): Habitat preferences of Birds in the Peryar Tiger reserve, Kerala. Ind. J. Forestry, 12; 288 - 295.

2013, Impact of Global Climate Change on Earth Ecosystems Pages 241–246
Editors: D.R. Khanna, A.K. Chopra, Gagan Matta, Vikas Singh & Rakesh Bhutiani
Published by: BIOTECH BOOKS, NEW DELHI

Chapter 29

Ex-situ Conservation of Rare, Endangered and Medicinal Plants

S.R. Kshirsagar

Post Graduate Department of Botany, S.S.V.P.S.L.K.Dr.P.R.Ghogrey, Science College, Dhule – 424 005, Maharashtra

Present paper communicates *ex-situ* conservation (in botanical garden) of rare, endangered and medicinal plants especially from areas of Khandesh and adjoining forest areas of Nashik district and South Gujarat

Introduction

Khandesh is northern part of Maharashtra State of India and divided into two districts in 1906 as East Khandesh (Jalgaon) and West Khandesh (Dhule and Nandurbar). Jalgaon is lies between 20° and 21°north latitude and 74°55′ and 76°28′ east longitudes, and Dhule and Nandurbar between meridians of longitudes 73°47′ and 75°11′ east and between the parallels of latitudes 20°38′ and 22°3′ north.

Forest areas of Khandesh includes Satpuras mountain and offshoots of Western Ghats (Sahyadris) and Western Ghats is one of the major hotspots of biodiversity in world. Forests of Khandesh is of dry, deciduous type with *Boswellia serrata, Hardwickia binata, Tectona grandis* and *Terminalia tomentosa* are dominant species apart from scrub types.

Ex-situ conservation of the taxa reported as rare, endangered and medicinal is underway in our five year (2010-2015) UGC, New Delhi, programme sanctioned as "PG Diploma in conservation of medicinal, endemic and endangered plants" under innovative programme of teaching and research in emerging areas

Flora of Jalgaon district includes 105 families of flowering plants with 442 genera and 813 species (Kshirsagar and Patil 2008) and Flora of Dhule and Nandurbar district includes 1133 species with 620 genera and 134 families (Patil 2003).

Methodology

Frequent field tours are arranged for survey and collection of seeds and saplings of rare, endangered and medicinal plants especially in Khandesh areas of North Maharashtra. Seeds are successfully germinated in nursery. Collected saplings are well in nursery.

Collected plant species are identified with the help of floras (Kshirsagar&Patil 2008, Cooke, 1958, Hooker 1872, Shah, 1978 etc., Shetty,&Singh 1987-1993).

Following plants species are conserved and multiplied in our botanical Garden and are arranged alphabetically by their botanical names followed by families in parantheses then it is follwed by local names, taxonomic identity, flowering and fruiting, localities, status and threats if any. Medicinal uses are appended at the end.

***Abrus precatorius* L. (Papilionaceae)**

Local Name: Gunj

Taxonomic Identity: Prostrate, trailing shrubs with white flowers

Flowering and Fruiting: More or less throughout the year

Locality: Manudevi

Present status: Rare

Threats: Huge collection for leaves and fruits by crude methods

Medicinal uses: Leaves are medicinal in stomatitis

***Amorphophallus commutatus* (Schott.) Engl.(Araceae)**

Local Name: Suran

Taxonomic Identity: Tall herbs with whitish flowers and red fruits *Flowering and Fruiting:* June- December

Locality: Manudevi

Status: Very rare

Medicinal Uses: Tubers are eaten and also used as vegetable and medicinally important in piles

***Aristolochia bracteata* Retz. (Aristolochiaceae)**

Local Name: Kidamar

Taxonomic Identity: Twining herbs with violet spotted large flowers

Flowering and Fruiting: August- November

Locality: Dhule and environs

Medicinal Uses: Juice of the plant is anthelmintic and used to kill worms and hence the local name Kidamar

Asparagus racemosus Willd. (Liliaceae)

Family: Liliaceae

Local Name: Shatawari

Taxonomic Identity: Twining shrubby plants with white flowers

Flowering and Fruiting: December- April

Locality: Pal

Present status: Rare

Threats: Huge exploitation for roots by natives

Medicinal uses: A reputed medicine "shatavari churna" is prepared from roots

Caralluma adscendens (Roxb.) R. Br.var. *fimbriata* (Wall.) Grav.& Mayur (Asclepiadaceae)

Local Name: 'Makadsing'

Taxonomic Identity: Succulent herbs with chocolate-red flowers

Flowering and Fruiting: August - December

Locality: Laling forest

Present Status: Very rare

Habitat: Found on rocky hills, boulders in association with *Euphorbia* antiquorum

Threats: Everyday exploited by local people as it is good vegetable

Medicinal uses: Medicinally important in mental balance

Centella asiatica (L.) Urban (Umbelliferae)

Local Name: Brahmi

Taxonomic Identity: Leaves orbicular with crenate margin

Flowering and Fruiting: November-February

Locality: Nakana

Medicinal Uses: Leaves are medicinal as brain tonic

Coccoloba uvifera (Polygonaceae)

Taxonomic Identity: Perennial shrubby plant leaves evergreen with ochreate stipules

Flowering and Fruiting: More or less throughout the year

Presents status in botanical garden: Nave of Central America but planted as evergreen leaves and very rarely grown in Indian Botanical Gardens

Cochlospermum religiosum (L.) Alst. (Cochlospermaceae)

Local Name: Jangali Kapas

Taxonomic Identity: Tall trees with dehiscent capsules having cotton within.

Flowering and Fruiting: February-June

Locality: Manudevi

Status: Very rare

Curcuma inodora Blatt. (Zingiberaceae)

Local Name: Jangali Haldi

Taxonomic Identity: Small herbs with pinkish flowers

Flowering and Fruiting: June- November

Locality: Manudevi

Status: Very rare

Uses: Rhizomes are used as vegetable

Dioscorea bulbifera L. (Dioscoreaceae)

Local Name: Kand

Taxonomic Identity: Twining herbs with creamy flowers

Flowering and Fruiting: July-November

Locality: Patnadevi

Present status: Occasional

Medicinal uses: Bulbs eaten raw and medicinal in piles

Eulophia herbacea Lindl. (Orchidaceae)

Local Name: Kukud Kand

Taxonomic Identity: Perennial herbs with fibrous tubers

Flowering and Fruiting: More or less throughout the year

Locality: Gujarat

Present status: Endangered in forests

Threats: Huge exploitation for tubers

Gloriosa superba L. (Liliaceae)

Local Name: Bachnag, Kallavi

Taxonomic Identity: Twining herbs with red-yellow flowers

Flowering and Fruiting: July-November

Locality: Patnadevi

Present status: Endangered

Threats: Highly exploited for rhizomes, flowers and seeds

Medicinal Uses: Colchicine –a poisonous alkaloid obtained from rhizome and is used in polyploidy

Ipomoea speciosa L.(Convolvulaceae)

Local Name: Samudrasok

Taxonomic Identity: Twining herbs with large violet flowers

Flowering and Fruiting: June- December

Locality: Gujarat

Medicinal Uses: Warm coconut oil applies on lower side of the leaves and tie over swellings to remove pus

Notonia grandiflora DC. (Compositae)

Family: Local Name: Wanar Roti

Taxonomic Identity: Herbs with glaucous leaves. Flowers yellow

Flowering and Fruiting: August-November

Locality: Patnadevi

Present status: Very Rare

Pedalium murex L. (Pedaliaceae)

Local Name: Motha Gokhru

Taxonomic Identity: Herbs with dark yellow flowers

Flowering and Fruiting: August-October

Locality: Dhule and environs

Present status: Rare

Threats: It is collected by herbalist as it is medicinal

Medicinal uses: Juice of the whole plant is used as coagulant

Rauwolfia tetraphylla L. (Apocynaceae)

Local Name: Sarpgandha

Taxonomic Identity: Shrubby plant with red fruits

Flowering and fruiting: More or less throughout the year

Medicinal uses: Reputed drug in cancer

Terminalia bellirica (Gaertn.)Roxb. (Combretaceae)

Local Name: Behda

Taxonomic Identity: Tall trees with creamy flowers

Flowering and Fruiting: February-June

Locality: Manudevi

Present status: Rare

Medicinal uses: Medicinally important as laxative

Withania coagulans (Solanaceae)

Family: Local Name: Panirbandh

Taxonomic Identity: Perennial suffrutescent herbs with creamy flowers

Flowering and Fruiting: December-February

Locality: Jaisalmer

Present status: Endangered in Jaisalmer of Rajashthan

Threats: Habitat loss and crude collection method as direct uprooting

Medicinal uses: Traditionally highly medicinal in coagulating milk into curd and hence the name panirbandh

Acknowledgements

I am thankful to UGC, New Delhi for sanctioning the programme to extend such work because conservation is the need of the days

References

Cooke, T.(1958 Repr. ed.) Flora of Presidency of Bombay, Vol.I-III, Botanical Survey of India, Calcutta

Hooker, J.D.(1872-1897) The Flora of British India, Vol. I-VII, L. Reeve and Co.London,U.K.

Kshirsagar S. R. and Patil, D. A. (2008) *Flora of Jalgaon District*, Maharashtra, M/S Bishen Singh Mahendra Pal Singh, Publishers and Distributors of Scientific Books, Dehradun, India.

Lakshminarasimhan and Sharma, B.D. (1991) *Flora of Nashik district*, Maharashtra, Botanical Survey of India, Calcutta.

Patil, D.A. (2003) *Flora of Dhule and Nandurbar Districts*, Maharashtra, Bishen Singh Mahendra Pal Singh, Dehra dun

Shetty, B.V. and Singh V. (1987-1993) *Flora of Rajasthan*, Flora of India, Series: 2, Vol.I-III, Botanical Survey of India, Calcutta.

2013, Impact of Global Climate Change on Earth Ecosystems *Pages* ***247–250***
Editors: **D.R. Khanna, A.K. Chopra, Gagan Matta, Vikas Singh & Rakesh Bhutiani**
Published by: **BIOTECH BOOKS, NEW DELHI**

Chapter 30

Natural Dyes: A Source of Medicinal Importance and Alternatives to Save Enviroment

Hemlata A. Padole and Neeta A. Tiwade
Department of Textile and Clothing
Sevadal Mahila Mahavidyalaya, Nagpur, Maharashtra

The Worldwide demand of natural dyes is nowadays of great interest due to the increased awareness on therapeutic properties of natural dyes in public. The history of medicine dates back to the origin of the human race. In India, the reference to the curative properties of some herbs in the Rig Veda seems to be the earliest records of the use of plants in medicine. India has long history of the use of large number of medicinal and aromatic plants for various purposes other than there medicinal use such as dyeing, printing, extraction of fibers, aroma purposes, used as spices, making household items etc. The impact of the textile industry on the environment and the consumption of raw materials and natural resources are becoming prime concerns. Natural dyes are derived from naturally occurring sources such as plants, insects, animals and minerals. Due to the German ban on synthetic dyes because of their carcinogenic, pollutative and non-biodegradable nature; the attention has shifted to the use of natural dyes. Among the all natural dyes, plant-based pigments have wide range of medicinal values. Although known for a long time for dyeing as well as medicinal properties, the structures and protective properties of natural dyes have been recognized only in the recent past. This study describes the detail information about natural dye yielding sources, which are helpful to further development of study.

Keywords: *Dyes, Medicinal value, Environment, Natural dyes.*

Introduction

Natural dyes are known for their use in colouring of food substrate, leather as well as natural protein fibers like wool, silk and cotton as major areas of application since pre-historic times. Until the middle of last century, most of the dyes were derived from plants or animal sources by long and elaborate process. Among these Indigo, Tyrian purple, Alizarin, Cochineal and Logwood dyes deserve special mention.

The use of non-allergic, non-toxic and ecofriendly natural dyes on textile has become a matter of significant importance due to the increased environmental awareness in order to avoid some hazardous synthetic dyes. However, worldwide the use of natural dyes for the colouration of textiles has mainly been confined to artisan/craftsman, small scale/cottage level dyers and printers as well as to small scale exporters and procedures dealing with high-valued ecofriendly textile production and sales. Recently, a number of commercial dyers and small textile export houses have started looking at possibilities of using natural dyes for regular basis dyeing and printing of textiles to overcome environmental pollution caused by the synthetic dyes. Natural dye produces very uncommon, soothing and soft shades as compared to synthetic dyes. On the other hand, synthetic dyes are widely available at an economical price and produce a wide variety of colours; these dyes however produce skin allergy, toxic wastes and other harmfulness to human body.

In spite of the better performance of synthetic dyes, recently the use of natural dyes on textile materials has been attracting more and more scientists for study on this due to the following reasons:

- Wide viability of natural dyes and their huge potential.
- Availability of experimental evidence for allergic and toxic effects of some synthetic dyes, and non toxic and non-allergic effects of natural dyes.
- To protect the ancient and traditional dyeing technology generating livelihood of poor artisan/dyers, with potential employment generation facility.
- To generate sustainable employment and income for the weaker section of population in rural and sub-urban areas both of dyeing as well as for non-food crop farming to produce plants such natural dyes.
- To study the ancient dyeing methods, coloured museum textiles and other textiles recorded by archaeology for conservation and restoration of heritage of old textiles.
- Specialty colours and effects of natural dyes produced by craftsman and artisans for their exclusive technique and specialty work.
- Availability of scientific information on chemical characterizations of different natural colorants, including their purification and extraction.
- Availability of knowledge base and database on application of natural dyes on different textiles.

Need of Using the Natural Dyes

National and international awareness about depletion of natural resources, ecological imbalance, pollution problem and our disturbed environment due to the ample usage of hazardous chemicals and particularly synthetic dyes have forced us to think of safer alternative. These factors have brightened the scope of utilization of natural dyes. Synthetic dyes are not only toxic and harmful to human skin but also to our environment. So a ban has been imposed on their use by many European countries. This implies the need to reintroduce the use of natural dyes for dying textiles. Natural dyes are biodegradable and eco-friendly. Growing plants for extracting natural dyes also leads to encouragement to forestation to which ultimately results in greater ecological balance. They are safe and compatible with environment. They also have medicinal and curative properties. Natural dyes are soft in color, cool to eyes and good to skin.

Medicinal Properties of Natural Dyes

Nowadays, herbal wear is also gaining a lot of importance due to its availability at cheaper rate and as garments used as medicine because of their medicinal properties. Generally, garments dyed with natural elements or plants, roots, seeds, flowers, leaves are called herbal wear. Herbal wear has medicinal properties:

- ☆ It is anti-allergic.
- ☆ It is anti-microbial.
- ☆ It has antiseptic properties.
- ☆ Such garments have a good breathability.
- ☆ The fabric is also eco-friendly as the waste from such herbal dyeing can be transformed into manure.
- ☆ It provides pollution free environment

Advantages of Natural Dyes

- ☆ Obtain from renewable sources
- ☆ No health hazards, sometimes they act as health cure
- ☆ Practically very mild chemical reactions are involved
- ☆ No disposal problems
- ☆ They are un sophisticated and harmonized with nature
- ☆ They help to provide employment for the rural people and preserve our traditional craftsmanship.
- ☆ They are safe and compatible with environment
- ☆ They have medicinal and curative properties
- ☆ They are soft in color, cool to eyes and good to skin.
- ☆ They have higher tinctorial value
- ☆ They are ecofriendly
- ☆ They have better biodegradability

- ✰ They have higher compatibility with environment
- ✰ It saves energy because of raw materials are not from petroleum products
- ✰ Waste material is used for the manufacture of compost

Disadvantages of Natural Dyes

- ✰ Availability
- ✰ Colour yield
- ✰ Complexity in dyeing process
- ✰ Reproducibility of shade

Future of Natural Dyes

Several anthraquinones compounds have been identified as secondary metabolites produced by fungi; such principles may be investigated and extended to evolve a biotechnological solution combining the advantage of both natural and synthetic dyes. The lost cultivation technique for the staple vegetable coloring matters must be researched in ordered to achieve acceptable shade gamut and fastness properties. Natural disperse dyes must be developed for manmade fibers. In the non toxic areas like food, drugs and cosmetic natural dyes alone can be successfully used. It will be the more fascinating carrier, if the proper attention is given natural dyes has a bright future further now in is suppressed due to the lack of knowledge about that.

Conclusion

Natural dyes are not only having dyeing property but also having the wide range of medicinal properties. Nowadays, fortunately, there is increasing awareness among people towards natural dyes and dye yielding plants. Due to their non-toxic properties, less side effects, more medicinal values, natural dyes are used day-to-day food products and in pharmaceutical industry. More detailed studies and scientific investigations are needed to assess the real potential and availability of natural dye yielding resources in great demand on the therapeutic formulations of natural drugs commercially.

References

Arora A. (2009), Textbook of Dyes

Chengaiah B. *et al*. (2010), Medicinal importance of natural dyes, Vol.2

Samanta A.& Agarwal p.(2009) Application of natural dyes on textiles, IJFTR journal Vol.34

Vankar P.(2000), Chemistry of Natural Dyes

Vankar P.(), Handbook on natural dyes for industrial applications

2013, Impact of Global Climate Change on Earth Ecosystems *Pages 251–256*
Editors: D.R. Khanna, A.K. Chopra, Gagan Matta, Vikas Singh & Rakesh Bhutiani
Published by: BIOTECH BOOKS, NEW DELHI

Chapter 31

Antagonistic Activity of Chitinase Enzyme and Effect Against Phytopathogens

Parikshit A. Bhamburkar and Ashwini C. Makhale
Modern College of Arts, Science and Commerce, Ganeshkhind, Pune – 53, Maharashtra

Serratia marcescens was isolated from soil. It is a Gram negative rod shaped bacterium, family Enterobacteriaceae. The growth of *Serratia marcescens* on potato dextrose agar (pH = 5) and nutrient agar (pH = 6.5) supplemented with 1 per cent chitin and the production of clear zone around the bacterial colony indicated the antagonistic activity of the chitinase enzyme produced by the bacterium. The antagonistic activity of the chitinase enzyme was tested against mainly two fungi such as *Aspergillus niger* and *Curvularia lunata*. *Serratia marcescens* was grown in a medium containing 1 per cent colloidal chitin. Whatman filter paper discs were soaked in *Serratia marcescens* broth and inoculated on the plates grown with fungi. Colloidal chitin was used as the substrate for the production of the enzyme chitinase. Chitinolytic microorganisms have been suggested for the control of fungi. Some mechanisms are involved in the antagonism and synergistic action of several enzymes such as glucanases, lipases, proteases, and chitinases.

Interest in biological control of plant pathogens has increased in recent years fuelled by trends in agriculture towards greater sustainability and public concerns over the use of hazardous pesticides in the environment. Most studies on the biological control of fungal plant pathogens focus on the use of antagonistic rhizobacterial strains belonging to the genus *Pseudomonas* or *Bacillus*. The development of bio control products based on isolates belonging to the genus *Serratia* is now gaining momentum.

Keywords: *Serratia marcescens, Chitinase, Antagonistic activity, Biocontrol.*

Introduction

Chitinases, which hydrolyse chitin, are present in a wide range of organisms including viruses, bactèria, fungi, insects, higher plants, and mammals. Most organisms (bacteria, plants and insects) have large families of chitinases with distinct functions, including digestion, cuticle turnover, and cell differentiation.Based on amino acid sequence similarity chitinases can be grouped into glycosyl hydrolase families (GH) 18 and 19, which are structurally unrelated. Bacterial chitinases belong to both family GH18 and 19.Recently; chitinases have gained interest in different biotechnological applications due to their ability to degrade chitin in the fungal cell wall and insect exoskeleton, leading to their use as antimicrobial or insecticidal agents. Another interesting application of chitinase is for bioconversion of chitin, a cheap biomaterial, into pharmacological active products, namely *N*-acetyl glucosamine and chito-oligosaccharides.

Chitin is one of the most abundant renewable biopolymer on earth that can be obtained as a cheap renewable biopolymer from marine sources. It is biocompatible, biodegradable and bio-absorbable, with antibacterial and wound-healing abilities and low immunogenicity; therefore there have been many reports on its biomedical application, food technology, material science, microbiology, agriculture, wastewater treatment, drug delivery systems, tissue engineering, bio nanotechnology. Chitin, a poly-beta-1,4-*N*-acetylglucosamine (GlcNAc) and is the main component of arthropod exoskeletons, tendons, and the linings of their respiratory, excretory, and digestive systems. Chitin is also an important component of the cell wall of fungi. Unlike cellulose, chitin can be a source of nitrogen as well as carbon (C: N = 8:1)

The bacterial species used for the current study is *Serratia marcescens*.Serratia marcescens is a species of Gram- negative, rod shaped bacterium in the family Enterobacteriaceae. Itis a motile organism and can grow in temperatures ranging from 5–40°C and in pH levels ranging from 5 to 9. It is differentiated from other Gram-negative bacteria by its ability to perform casein hydrolysis.

Pathogenic micro organisms affecting plant health are a major and chronic threat to food production and ecosystem stability worldwide. As agricultural production has intensified over the past few decades, producers have become more dependent on agrochemicals as a relatively reliable method of crop protection helping with economic stability.However increasing use of chemical inputs causes several negative effects such as development of pathogen resistance to the applied agents and the non – target environmental impacts. There are also a number of fastidious diseases for which chemical solutions are few. Biological control is thus being considered as an alternative way of reducing the use of chemicals in agriculture.

Potential uses of naturally occurring bacteria, actinomycetes and fungi replacement or supplements for chemical pesticides have been addressed in many studies. Biological control of damping-off in crops caused by *Rhizoctonia solani*has been reported all over the world using application of antagonistic fungi and bacteria isolated from coastal soils.Chitinolytic microorganisms have been suggested for the control of some fungi. *Serratia marcescens*chitinase reduced the incidence of disease in bean seeds caused by *Sclerotiumrolfsii; Aeromonascaviae*controlled the infection caused by *Rhizoctoniasolani*and *Fusariumoxysporum*in cotton. Chitinase from different *Actinomycetes*has also produced inhibition of phytopathogenic fungi in cotton, rice, soybean,and tomato. The control agent *Trichoderma*is well known for having a broad antagonism against different fungi.

Materials and Methods

Microorganisms

1. *Serratia marcescens* culture was maintained on nutrient agar slants. Loopful of this culture was inoculated in 100 ml of nutrient broth containing chitin and incubated at 37°C for 24 hours. This was then used as the seed culture for the inoculation of broth for chitinase enzyme production.
2. *Aspergillus niger* and*Curvularia lunata*cultures were maintained on potato dextrose agar slants. Loopful of this culture was inoculated in 100 ml of potato dextrose broth and incubated at room temperature for 72 hours.

Media Composition

Plates for testing antagonistic activity were made with the use of Potato Dextrose Agar (Hi Media) and Nutrient Agar with 1 per cent colloidal chitin. Each medium contained 2 per cent agar (w/v).The pH of PDA medium was adjusted to 5.0 and the pH of nutrient agar chitin was adjusted to pH 6.5 using 1N NaOH/1N HCl. The medium was autoclaved at 121°C at 15 psi for 20 minutes.

Preparation of Colloidal Chitin

Colloidal chitin is the product obtained from the Chitin powder. It is used as the substrate for the production of the enzyme Chitinase. Chitin flakes are insoluble in water. Colloidal chitin is prepared by acid hydrolysis of grinded chitin flakes. 10gm chitin dissolved in concentrated HCl. Stir overnight at 4°C. Add ice cold ethanol and stir overnight at room temperature. Centrifuge contents at 10,000rpm for 10 mins and collect precipitate. Neutralize the precipitate with 0.1 M sodium phosphate buffer. Filter colloidal chitin using Whatmann filter paper. Store at 4°C. (Roberts and Selitrennikoff, 1988)

Chitinase Enzyme Assay

Chitiase assay was determined spectrophotometrically (Miller *et al.*) 0.5ml of culture supernatant is added and mixed to 1ml of 1 per cent colloidal chitin in 0.1 M sodium phosphate buffer (pH=7).The reaction is carried out at 37°C for 30 minutes. Centrifuge at 12,000 rpm for 10 minutes and decant the sample. Add 1 ml of DNSA reagent and boil in water bath. Read absorbance at 540nm. One enzyme unit is defined as the amount of enzyme that produces 1 μmol reducing sugar per hour under the reaction conditions.

Protocol for Antagonistic Activity

Prepare 100ml of potato dextrose agar (pH=5) and nutrient agar (pH=6.5).Inoculate loopful of fungal culture at one corner of the petri plate containing medium. Prepare autoclaved Whatman filter paper discs of 5 mm each. Dip the paper discs in *Serratia marcescens* culture broth and soak for 5 minutes. Place these filter paper discs on the other side of the petri plates having fungal inoculum. Incubate at room temperature and observe for the inhibition of fungal growth by chitinase enzyme.

Observations

Antagonistic activity of the chitinase enzyme of *Serratia marcescens* against the fungus *Curvularia lunata* was observed by the presence of a clear zone around the bacterial disc (Figures 31.1 and 31.2).

Conclusions

- ☆ By observing the clear zone around the colony of the bacterium *Serratia marcescens*on chitin supplemented media it can be concluded that the bacteria is Chitinolytic and produces the enzyme chitinase
- ☆ The presence of the clear zone indicates hydrolysis of chitin in the medium by chitinase enzyme secreted extracellulary by *Serratia marcescens.*

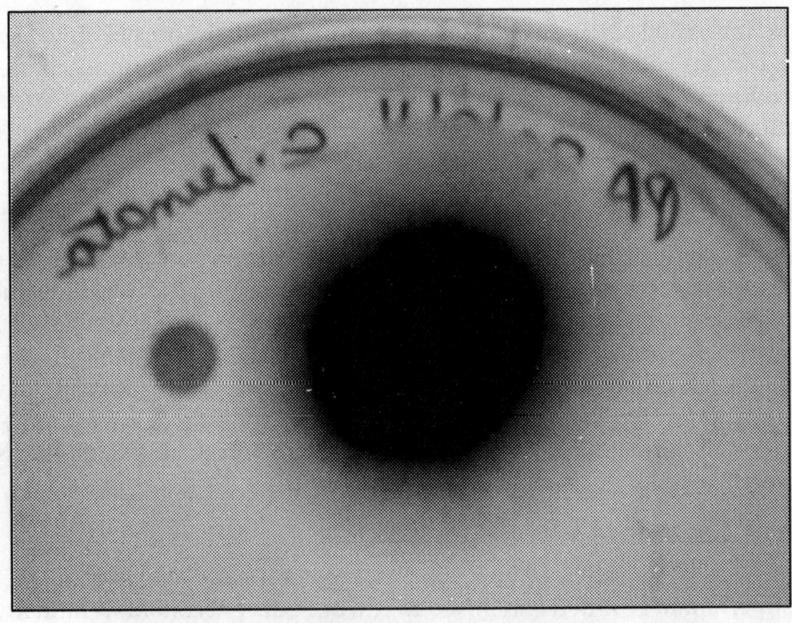

Figure 31.1: Antagonistic Activity of Chitinase Enzyme of *Serratia marcescens* against *C. lunata*

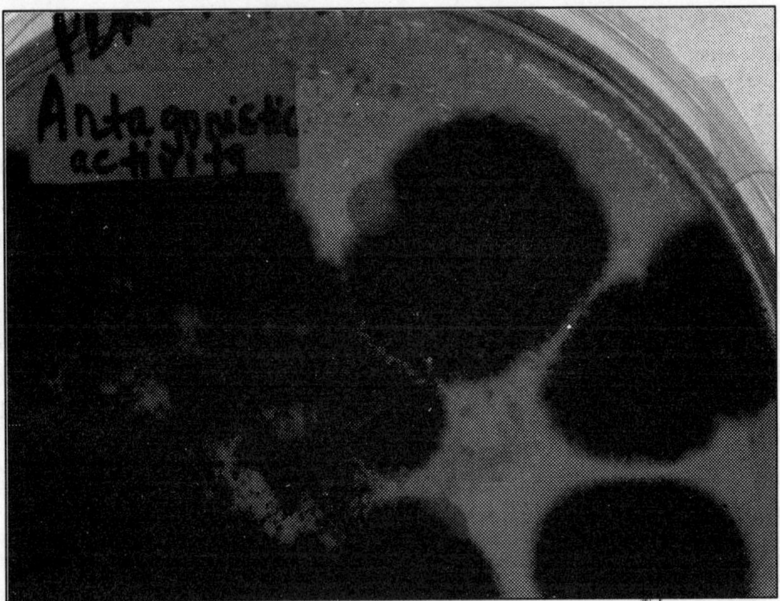

Figure 31.2: Antagonistic Activity of Chitinase Enzyme of *Serratia marcescens* against *A. niger*

- ☆ The growth of fungi is restricted by the chitinase produced by the bacterium and this reveals that bacterial chitinase has antagonistic activity on the fungal pathogen and thus can inhibit the phytopathogen.

Acknowledgement

I sincerely thank Dr.RekhaGupta(HOD,Dept. of Biotechnology, Modern College Ganeshkhind, Pune) and Miss AditiDeshpande for their timely help and guidance.

References

A.Reyes-Ramirez, B.I. Escudero- Abarca, G. Aguilar-Uscanga, P.M. Hayward-Jones, and J. EleazarBarbozacorona.(2004) Antifungal activity of *Bacillus thuringiensis*chitinase and its potential for the biocontrol of phytopathogenic fungi in Soybean seeds. *Journal of Food Science*. Vol. 69, Nr 5, 2004

Chigaleichik, A.G.; Pirieva, D.A.; Rydkin, S.S.(1976) Chitinase from *Serratia marcescens* BKM B-851. *Prikl. Biokhim. Mikrobiol.*, *12*, 581–586.

David De Vleesschauwer and Monica Hofte. (2007) Using Serratia plymuthica to control fungal pathogens of plants. CAB Reviews: *Perspectives in Agriculture, Veterinary Science, Nutrition and Natural Resources* 2, No. 046

K.Parani and B.K. Saha (2009). Studies on interaction of *Serratia marcescens*Strain(SR1) with fungal pathogens. *American-Eurasian Journal of Agriculture and Environmental Science*, 5 (2), 215-218

Mauch F., Hadwiger L.A., and Boller T. (1988) Antifungal hydrolases in pea tissue. Purification and characterization of two chitinases and two ß-1,3-glucanases differentially regulated during development and in response to fungal infection. Plant Physiology 87, 325-333.

Monreal, J., and E. T. Reese. (1969). The chitinase of *Serratia marcescens*. Can. J. Microbiol. 15:689-696.

ORANUSIN,.A. &TRINCIA,.P. J. (1985). Growth of bacteria on chitin, fungal cell walls and fungalbiomass, and the effect of extracellular enzymes produced by these cultures on the antifungal activityof amphotericin. B. *Microbios*43, 17-30.

V. Gohel, C.Megha, P. Vyas and H.S. Chhatpar. (2004) Strain improvement of Chitinolytic enzyme producing isolate *Pantoeadispersa* for enhancing its biocontrol potential against fungal plant pathogens. *Annuals of Microbiology*, 54 (4), 503-515.

Walden K. Roberts,Claude P. Selitrennikoff.(1988)Plant and Bacterial Chitinases Differ in Antifungal Activity. In *Journal of' General Microbiology*, 134, 169- 1 76.

Z. Kamil, M. Rizk, M. Saleh and S. Moustafa. (2007). Isolation and identification of Rhizosphere soil Chitinolytic bacteria and their potential in Antifungal Biocontrol. Global Journal of Molecular Sciences 2 (2), 57- 66.

2013, Impact of Global Climate Change on Earth Ecosystems *Pages* ***257–262***
Editors: **D.R. Khanna, A.K. Chopra, Gagan Matta, Vikas Singh & Rakesh Bhutiani**
Published by: **BIOTECH BOOKS, NEW DELHI**

Chapter 32

Increasing Pollution, Raising Health Problems, and Omega-3 Rich Diet

Mansi M. Sharma

Modern College of Arts, Science and Commerce, Ganeshkhind, Pune – 5, Maharashtra

Omega-3 fatty acid helps to promote and accelerate the better health development in chick embryo: proved experimentally. It also helps to maintain the developmental patterns in the embryo. Rising pollution has mutagens and such compounds which are responsible for production of free radicals. The Ω-3 fatty acid produces prostaglandins in the body, which are important to reduce inflammation in the body. The Ω-3 fatty acid is an essential fatty acid for our body. This plays an important role in maintaining ourselves fit. This group of fatty acids has 3 main fatty acids like ALA, EPA and DHA. As we all know, DHA is important for brain development. Ω-3 fatty acids are known to carry out many functions as reduction of inflammation, lowering the cholesterol level, prevention of cancer, maintaining fluidity of cell membrane and reducing the risk of cardiovascular disease. In normal conditions In normal cells, *BRCA1* and *BRCA2* help ensure the stability of the cell's genetic material (DNA) and help prevent uncontrolled cell growth.

Rise in pollution is killing us in all ways. Drinking pure water and breathing fresh air, both are becoming difficult day by day. All these pollutants leads to a disaster called as free radicals. These radicals produce oxidative stress in the body.

Air pollution is the cause of many health problems such as asthma, chronic bronchitis, heard attack and even cancer. On the other hand, the drinking water is also contaminated with around 50,000 chemicals. Discharge of herbicides and pesticides in water increases the risk of health problems. The air and water carries so many mutagens with it, which are responsible for cancer.

Keywords: *Pollution, Ω-3 fatty acids, Free radicals, Mutagens.*

Introduction

Cancer is nothing but excessive and uncontrolled growth of cells which result in tumor formation.When the cell mutate it goes on multiplying and don't undergo PCD. Not all tumors are cancerous; tumors can be benign or malignant. Treatment of cancer includes chemotherapy, radiation, surgery, hormone therapy. *BRCA1* and *BRCA2* are human genes that belong to a class of genes known as tumor suppressors, inactivation or mutation of these genes causes development of breast and ovarian cancer. Anticancer drugs work on different targets, therefore they not only affect the cancerous cells but other cells as well.Omega-3 fatty acids is a family of unsaturated fatty acid, Nutritionally important omega 3 fatty acids are: ALA (α linolenic acid), EPA (eicosapentaenoic acid), DHA (docosahaexenoic acid). These acids are known to reduce risk of many diseases such as cancer. As well as omega-3 fatty acid is a major component of brain gray matter and of the retina in the most mammalian species. Cancer is known to be a fatal disease. Cure for cancer is not full-proof. Though many synthetic anticancer drugs have been discovered, they have numerous side effects such as organ damage which is irreversible. These anticancer drugs kill cancer cells and also damage other tissues in the body. On the other hand, omega 3 is a naturally occurring substance which is recently known to have an anticancer property. Since it is naturally occurring, its side effects will be negligible as compared to synthetically prepared anti cancer drugs. The work is done to observe the anticancer activity of omega 3.

Materials and Methods

Omega-3 fatty acid was taken from the commercially available capsules of it. The working solutions of 0.1ppm, 0.3ppm, 0.5ppm with sterile distilled water. These solutions were injected on blastodisc of the embryo using window technique. The eggs were sealed and incubated for different incubation periods. The isolated embryos were compared with respective controls.

In window technique the window is made in the shell of egg at the side of air space, the blunt surface. Then the internal membrane which is chorio allantoic membrane is removed carefully and then with the help of forceps. After that, around 200μl of working solution is slowly poured on the blastodisc such as it is spread equally on the blastodisc. Blastodisc is present in the vitelline membrane on yolk ball. These eggs are then sealed and incubated. After incubation the blastodisc were isolated carefully alongwith the part of vitelline membrane. The blastodiscs were carfully cleaned and washed to remove the yolk present on it. These embryos the placed on slide and observed under microscope on high power. The number of somites, the development of heart, development of neural tube, vitelline membrane and structure of area vasculosa and area pellucid is noted down.

Results

Results after 30 Hours

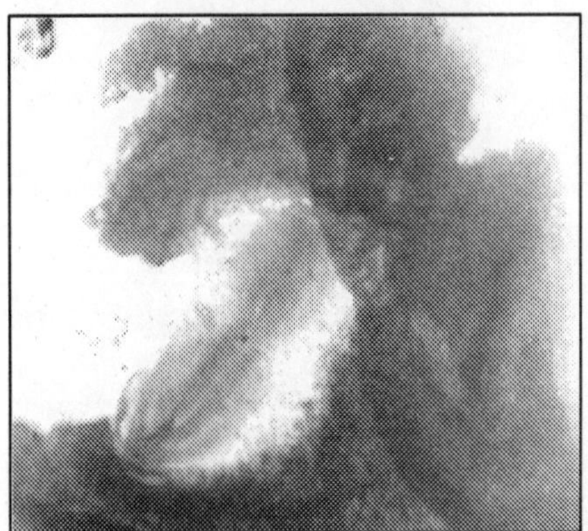

Figure 32.1: Treatment of Eggs with 0.1 ppm Conc. of Omega-3 Fatty Acid on 18 Hour Stage

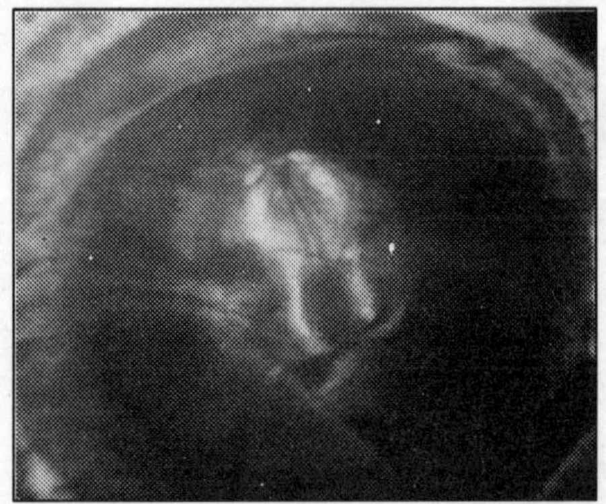

Figure 32.2: Treatment of Eggs with 0.3 ppm Conc. of Omega-3 Fatty Acid on 18 Hour Stage

Results after 43 to 48 Hours

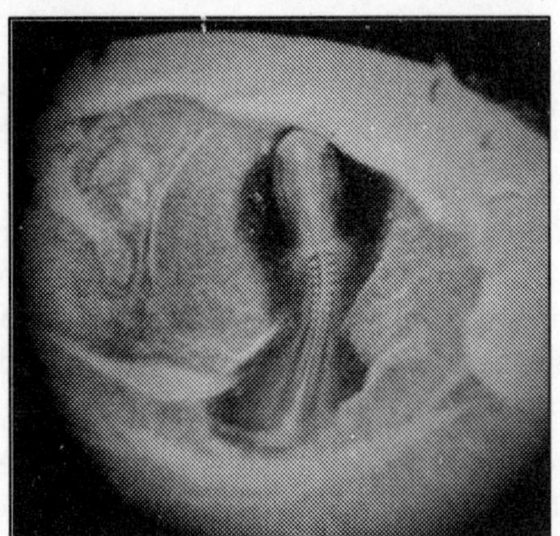

Figure 32.3: Treatment of Eggs with 0.1 ppm Conc. of Omega-3 Fatty Acid on 18 Hour Stage

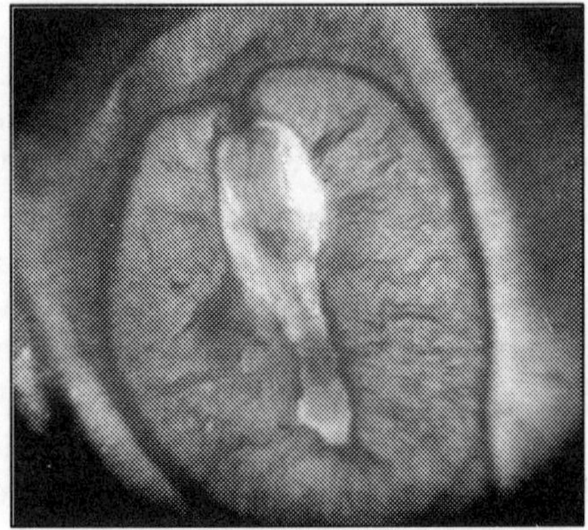

Figure 32.4: Treatment of Eggs with 0.3 ppm Conc. of Omega-3 Fatty Acid on 18 Hour Stage

Observation

When Figure 32.1 was compared with the control, it was seen that number of somites was 4. It was observed that neural tube and endoderm of foregut was well developed. It was also marked with presence of proamion, notochord and blood islands in the region of area opaca.

Results after 53 Hours

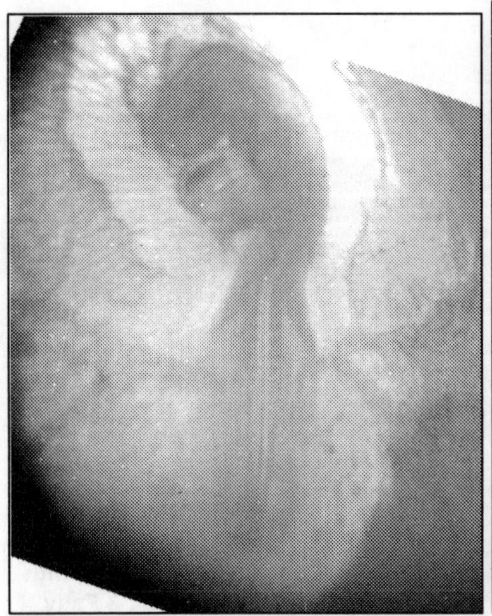

Figure 32.5: Treatment of Eggs with 0.1 ppm Conc. of Omega-3 Fatty Acid on 18 Hour Stage

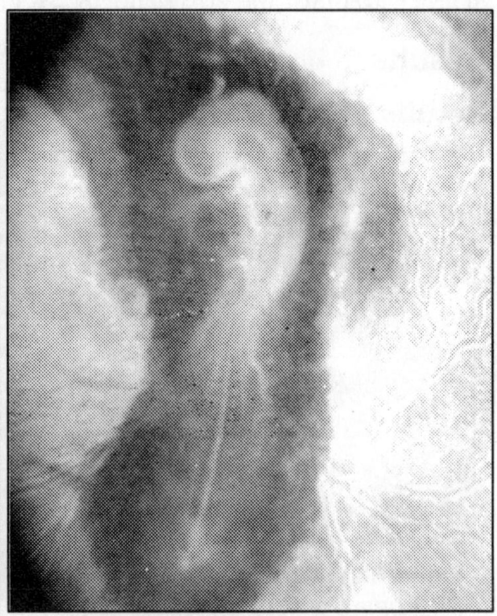

Figure 32.6: Treatment of Eggs with 0.3 ppm Conc. of Omega-3 Fatty Acid on 18 Hour Stage

Comparison of Figure 32.2 with control showed that anterior neuropore well marked and open, lateral horns of mesoderm and intersomitic furrows well developed, number of somite 6 and rhomboidal sinus was present.

It was seen that the embryo in Figure 32.3 was extremely well developed and also the somites were well developed with 10 in number. Brain vesicles were normal and vitelline veins were seen forming and normal in architecture. Primitive streak, primitive groove and hensen's node at the posterior most boundary of area pellucida. Also embryo showed all normal structures *i.e.* normal morphogenesis.

When Figure 32.4 was compared with control, the area vasculosa excessively developed and area pellucida seen restricted, lateral horns of mesoderm well developed and heart was present.

Figure 32.5 on comparison with control showed beating heart. Vitelline arteries well developed with proper development of midbrain and hindbrain. Blood is observed in region of area vasculosa.

When Figure 32.6 was compared with control, beating heart was seen and vitelline arteries well developed. Proper development of midbrain and hindbrain was seen. Blood is observed in region of area vasculosa and developing eye has been observed. Well developed network of blood vessels.

Conclusion

Since omega-3 fatty acids are growth promoters, all the embryos treated with omega-3 fatty acid showed accelerated growth as compared to the control embryos of same stage, but no abnormal growth *i.e.* cancerous growth. As EPA is important in decreasing inflammation, and DHA is important in the eye and brain, it can be included in our diet.

Also our cells produce destructive particles called free radicals as they go through their normal metabolic functions. Pollution is also the cause of production of free radicals. Also free radicals are produced in body during inflammation, stress, environmental toxins, ultraviolet radiation, x-ray radiation, low magnesium levels and immune activation. Free radicals are both a cause and a result of inflammation. Damage from free radicals causes inflammation; chronic inflammation goes on to produce lots of free radicals which in turn create more inflammation and research says that EPA helps to decrease inflammation. Our body can produce EPA and DHA when provided with ALA in the diet. The main reason to incorporate essential fatty acids in diet is, they can deactivate free radicals by giving them electrons. The fatty acids also sweep toxins along to the liver to be eliminated.

Acknowledgement

I sincerely thank my project guide Prof. Pinakin Wagh for his guidance stimulating discussions and valuable time. I am thankful to Dr. Rekha Gupta, Head, Department of Biotechnology, modern college, Ganeshkhind, Pune for providing me all the departmental facilities and guidance.

References

1. A SERIES OF NORMAL STAGES IN THE DEVELOPMENT OF THE CHICK EMBRYO VIKTOR HAMBURGER, *Department of Zoology, Washington University, St.* ***Louis,*** Nissouri, HOWARD L. HAMILTON*Department* ***of*** *Zoolo8gy and Entom2070gy, Iowa State College, Ames*
2. Chick Embryos in Shell-less Culture, *Cynthia J. Fisher,*
3. Culture of Chicken Embryos in Surrogate Eggshells,S. Borwornpinyo, J. Brake, P. EMozdziak,1 and J. N. Petitte, Department of Poultry Science, North Carolina State University, Raleigh, North Carolina 27695
4. DEVELOPMENTAL BIOLOGY, by Scott Gilbert, 10th edition
5. Fish Oil Supplements Linked to Lower Risk of Breast Cancer: Study EXECUTIVE HEALTH http://www.businessweek.com/lifestyle/content/healthday/640917.html
6. GENE 7 and 10, byB. Lewin
7. Human Guinea Pig Success For Omega 3 Cancer Treatment, by Kate Melville, http://www.scienceagogo.com/news/20051010003544data_trunc_sys.shtml
8. MOLECULAR BIOLOGY OF GENE, by Watson
9. More support omega-3 may protect against colorectal cancer, By Stephen Daniells,

22-Nov-2006. http://www.nutraingredients.com/Research/More-support-omega-3-may-protect-against-colorectal-cancer

10. Omega 3 Curbs Precancerous Growths in Those Prone to Bowel Cancer, Study Suggests, *ScienceDaily* (*Mar. 21, 2010*) http://www.sciencedaily.com/releases/2010/03/100317212648.htm
11. Omega-3 Fatty Acids Reduce Risk Of Advanced Prostate Cancer *ScienceDaily* (*Mar. 25, 2009*) – Omega-3 fatty acids appear protective against advanced prostate cancer, and this effect may be modified by a genetic variant in the COX-2 gene, according to a report in *Clinical Cancer Research*, a journal of the American Association for Cancer Research. http://www.sciencedaily.com/releases/2009/03/090324131444.htm
12. PRACTICAL BIOCHEMISTRY, by Plummer
13. PRINCIPLES OF BIOCHEMISTRY; LEHNINGER, by Nelson and Cox
14. VERTEBRATE ZOOLOGY, by Kotpal
15. VERTEBRATE ZOOLOGY, by Verma and Agarwal

Index

F

G

S

T

U

V

W